# EINSTEIN

# EINSTEIN

## The Man and His Mind

SELECTIONS FROM
THE BERGER COLLECTION
CHAPEL HILL,
NORTH CAROLINA

Gary S. Berger, MD, and Michael DiRuggiero

*This book is dedicated*
*to the memory of Albert Einstein.*

*All royalties will be*
*contributed to the Albert Einstein Archives*
*at the Hebrew University of Jerusalem.*

## TABLE OF CONTENTS

# ESSAYS



# 1890–1920



# 1921–1930

## *1931–1940*



## *1941–1950*



## *1951–1955*



## *BACK MATTER*

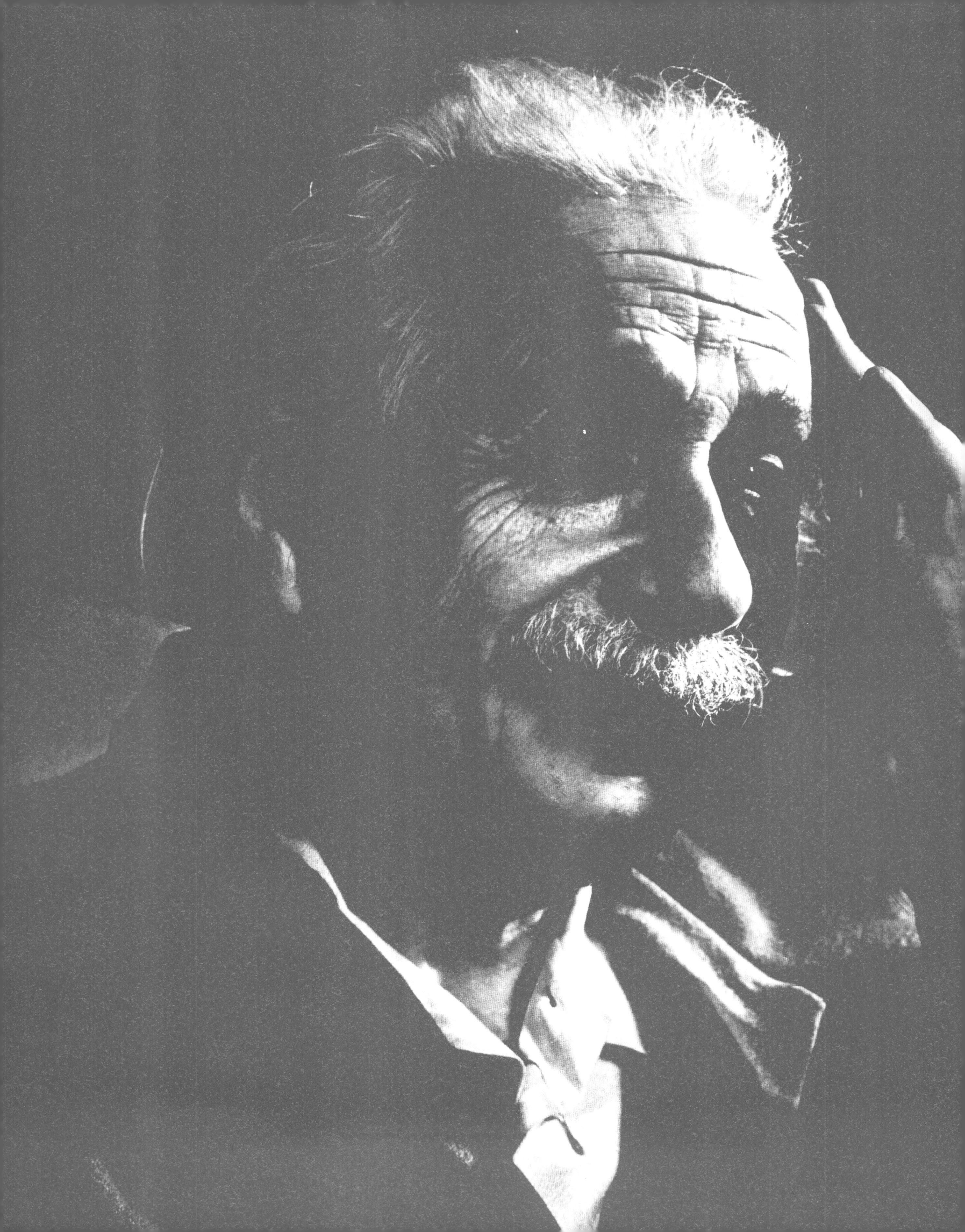

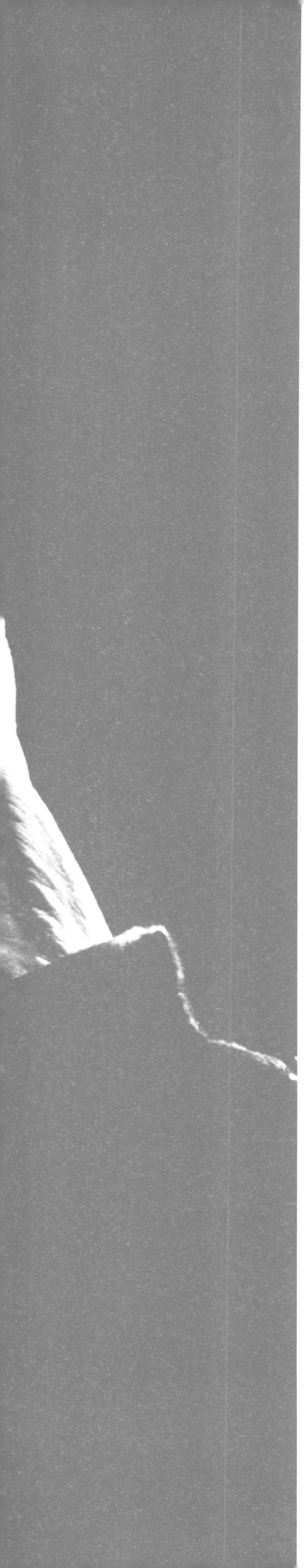

*Imagination is more
important than knowledge.
Knowledge is limited.
Imagination encircles
the world.”*

—ALBERT EINSTEIN, *1929*

## *PREFACE*

Albert Einstein's name was familiar to me in childhood, as it was to all my friends. It was synonymous with genius. We knew Einstein made important discoveries about the universe that had something to do with space and time.

When I first learned about special relativity in college, I was bewildered. Having other interests, I put the subject out of my mind. But it must have lingered, because years later it seemed important to revisit these strange concepts of time slowing down and lengths contracting.

I read and reread Einstein's paper *On the Electrodynamics of Moving Bodies*, which presented the special theory of relativity. I read discussions about that famous paper as well as articles and books by and about Einstein, and I watched all the videos about him that I could find. The more I learned, the more fascinated I became with how Einstein could recognize what no one else had before about the nature of the universe.

With insights from simple thought experiments, Einstein discovered that space and time are malleable and are shaped by matter. His general theory of relativity explained the force of gravity. It predicted the bending of light waves, black holes, gravitational waves, and the Big Bang. Einstein ushered in the era of quantum physics, having revealed light's dual wave/particle nature and the relationship between mass and energy. With thought and imagination, he transformed our understanding of the cosmos.

Eventually, I began collecting photographs and documents of the great scientist. Not being a physicist, I could appreciate his pictures, if not the complex mathematics in his writings. The photos gave me the feeling of a personal connection to Albert Einstein—the real, living man—almost as if I knew him.

In retrospect, there was an indirect connection. During weekends in my late teens, I frequented the home of a physicist, Max Herzberger, who had been Einstein's friend. On his living room wall was a beautiful portrait photograph of Einstein. Perhaps that image was the seed that led decades later to this collection.

As you explore these pages, I hope you enjoy the same feeling they have given me of affinity to one of the most extraordinary individuals who ever lived.

—GARY S. BERGER, MD

## *FOREWORD*

*Einstein: The Man and His Mind* is a unique and valuable addition to the avalanche of books that have been published on different aspects of the Einstein phenomenon. The authors bring to this endeavor many years of intellectual, emotional, and professional kinship with Albert Einstein—the man and the scientist. Dr. Gary S. Berger is an avid collector of documents related to Einstein. The backbone of his collection, and the backbone of the present book, is signed photographs of the physicist. Michael DiRuggiero, an active player in the vibrant market of Einstein-related archival material, has been helping to curate the Berger collection.

Albert Einstein was twenty-one years old at the beginning of the twentieth century, just graduated from the Federal Institute of Technology in Zurich (ETH) and in desperate search of an academic position that would allow him to pursue his ideas about the burning problems on the agenda of physics. About twenty years later he received the Nobel Prize, after having profoundly changed the scientific community's understanding of the physical world. At the end of the century he was elected by the readers of *Time* magazine as the "Person of the Century" (see page 158), followed by Mahatma Gandhi and Franklin Roosevelt.

The twentieth century is gone and its memories are receding into history, but the memory of its great hero still reverberates with ever growing intensity. Einstein was, first and foremost, the architect and engineer of a new understanding of the physical world, the most revolutionary innovator since Newton. He remains the uncontested pioneer of the transition from classical to modern physics.

The papers he published in 1905, his "miracle year," are the pillars of this transition. The technological consequences of these papers continue to affect our daily lives through myriad applications. In 1915, he formulated his masterpiece, a new theory of gravitation—the general theory of relativity—which became the genesis of modern cosmology and the basis of our understanding of the universe.

Einstein's fame stems from his fundamental, groundbreaking contributions to science. Yet they alone cannot explain his universal iconic status. One of the apparently unique aspects of the Einstein phenomenon is that his work left its mark on the cultural history of the twentieth century, far beyond his area of expertise. His presence in modern culture is all-encompassing—in art and literature, in movies and television programs, and in the digital media. He became one of the first media stars of science at a time when the world was eager to embrace such celebrities. Einstein was constantly in the public eye. In numerous articles and interviews, correspondence with peers, and public addresses, he expressed his views on a variety of political and moral issues—nationality and nationalism, war and peace, human liberty and dignity—and he launched relentless attacks on all forms of discrimination. Einstein's views and activities outside of physics were not simply add-ons to a life devoted to science. They were evidently driven by the same inner urge as his quest for scientific progress.

This humble dedication to human rights is probably the most profound reason why Einstein has become such a popular cultural icon; why centennials of the landmarks of his creativity have been celebrated worldwide with public events, international conferences, workshops, and television programs; why his image decorates so many commercial products and is the most recognizable face on our planet.

The present book contains numerous signed photographs from the Berger collection, illustrating chronologically different chapters in Einstein's life. They are interwoven with signed letters, Einstein quotations, covers of selected articles, and canonical equations, and are accompanied by descriptive remarks that are always informative, often thought-provoking, and sometimes amusing (for example, Einstein's meeting with Charlie Chaplin, page 94). All these annotated items draw a fascinating portrait of Albert Einstein, a genuinely modest person. The photographs from the second half of his life, in particular, evoke an image of a friendly non-conformist. The fame that was his fate did not corrupt him. He remained a simple man who was not striving to please anybody, by either action or behavior, words or looks. He was an eccentric who defied authority and convention, to the extent that he did not wear socks when visiting President Roosevelt in the White House. His hair was usually in disarray, and one six-year-old girl wrote to him, "Dear Mr. Einstein, I saw your picture in the newspaper. You ought to have your hair cut so that you can look better."

The authors' gracious gesture to contribute the royalties from this book to the Albert Einstein Archives at the Hebrew University of Jerusalem is in line with their decision to dedicate their book to Einstein's memory. It reflects their commitment to maintaining and spreading Einstein's legacy in the public eye.

The Albert Einstein Archives constitutes an extremely valuable historical resource. Considered one of the most significant sources for the history of modern physics, the Archives is additionally an extremely important source for German, European, Jewish, and American intellectual, political, and social history of the first half of the twentieth century. Its documents include materials relating to Einstein's lifelong scientific odyssey, and outside of physics there is also extensive documentation on pacifism, militarism, fascism, nationalism, McCarthyism, world government, and nuclear disarmament. Documents on antisemitism, the Holocaust, Israel, and the Arab-Israeli conflict are related to Einstein's Jewish identity. The Albert Einstein Archives also includes subcollections of non-textual materials: photographs, medals, honorary diplomas, Einstein trivia and collectibles, sound recordings, and film footage.

Einstein was one of the founding fathers of the Hebrew University of Jerusalem. Therefore, it was natural that he made it the eternal home of his intellectual legacy. The Archives, together with the Einstein Papers Project in Pasadena and with Princeton University Press, have produced a scientific publication of the documents in the archives and made it freely accessible online. The Archives has also shared this material with the general public by presenting selected documents from its holdings in various exhibitions and events related to Einstein's legacy worldwide. Therefore, the authors' gesture is certainly appropriate, and it is accepted with appreciation and gratitude.

—HANOCH GUTFREUND

*Professor (Emeritus) of Physics, The Hebrew University of Jerusalem, and Academic Head of the Albert Einstein Archives*

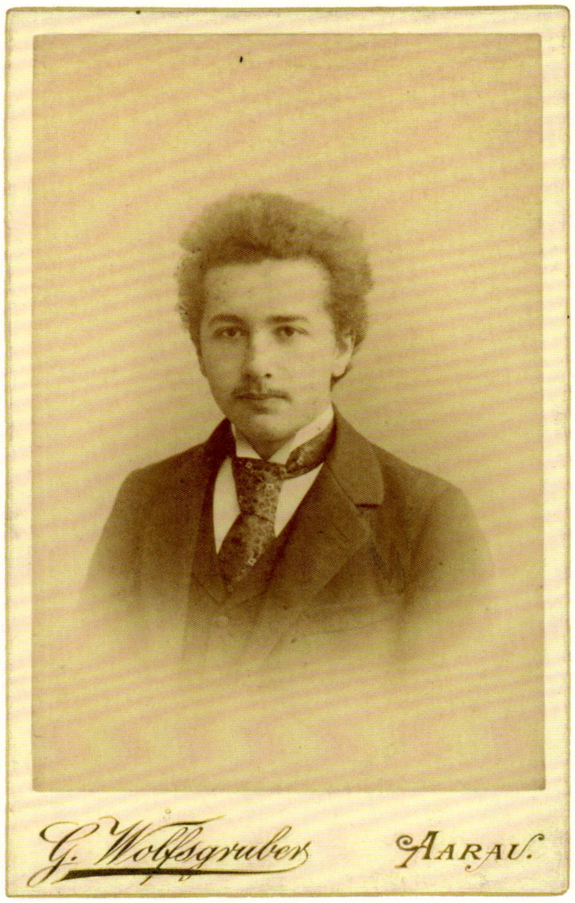
G. Wolfsgruber
AARAU.

## *LOVE AND LIGHT*

This is the earliest known signed photograph of Albert Einstein. We believe it was taken when he was seventeen years old to commemorate his graduation from the cantonal school of Aarau, Switzerland.

Einstein's year in Aarau was one of the happiest years of his life. At this school independent thought was encouraged, in sharp contrast to the authoritarian education he received in Germany, which he had hated. Einstein later recalled, "In Aarau I made my first rather childish experiments in thinking that had a direct bearing on the Special Theory. . . . If a person could run after a light wave with the same speed as light, you would have a wave arrangement which could be completely independent of time. Of course, such a thing is impossible."

The year 1896 was memorable for Einstein in other ways as well. While boarding with the Winteler family, he fell in love for the first time with eighteen-year-old Marie Winteler. And, to avoid mandatory German military service, which he detested, Einstein renounced his German citizenship (for the first time; he would renounce it for the second time in 1933). He was stateless for the next five years, finally becoming a Swiss citizen in 1901.

This photograph, a formal studio portrait in the carte-de-visite style, printed on card stock, was designed for presentation. Einstein gave it to his lifelong friend Albert Karr-Karusi and inscribed the back (in German): "To my dear Albert / Your Albert." It was a memento of their friendship, given in the spirit of today's high school students who sign each other's yearbooks.

Upon graduating with high marks from the Aarau secondary school, Einstein was admitted to the Polytechnic Institute of Zurich, where he embarked on his scientific education in earnest—and a most extraordinary life.

UNKNOWN PHOTOGRAPHER,
*Aarau, Switzerland, ca. 1896. Signed and inscribed by Einstein to Albert Karr-Karusi: "Meinem l[ieben] Albert / Dein Albert" ("To my dear Albert / Your Albert").*

## *EINSTEIN'S "MIRACLE YEAR": THE REVOLUTIONARY PAPERS OF 1905*

Albert Einstein published four papers in rapid succession at age twenty-six in the authoritative journal *Annalen der Physik* (the first three of which are in this collection, shown here). They shook the foundations of modern physics. He was working as a clerk third-class at the patent office in Bern, ironically having failed to obtain a position at a university.

In the first paper, Einstein accepted as real the quantization of radiation, conceived by Max Planck strictly as a mathematical tool. By proposing that light energy is carried as discrete quanta (later to be named photons), Einstein explained the mysterious photoelectric effect. His paper became a foundation of quantum theory. For this work, Einstein received the 1921 Nobel Prize in Physics.

In the second paper, Einstein explained Brownian motion. This paper is generally regarded as the first proof that molecules exist. Subsequent experiments confirmed Einstein's statistical derivation, providing solid evidence for the reality of atomic theory.

In the third paper, Einstein introduced the special theory of relativity. Einstein framed it as a "heuristic" argument, showing that if the speed of light is constant, and the laws of physics are the same for all frames of reference in uniform motion with respect to each other, then time and space are relative to the observer. Having overturned Isaac Newton's concept of absolute space and absolute time, Einstein expanded the theory to include accelerated motion in his general theory of relativity, which he published in 1916.

Einstein's fourth paper was a short derivation that he noticed as a consequence of his special relativity paper. It established the equivalence of mass and energy, leading to the most famous equation of twentieth-century science: $E=mc^2$.

Einstein's "miracle year" papers were so novel that they were met with silence at first, much to his disappointment. But soon, *Annalen der Physik*'s editor, Max Planck, sent his assistant to Bern to find out who this unknown author was. Einstein was on his way to being recognized, not just as a theoretical physicist deserving of an academic appointment, but as a scientist of exceptional genius.

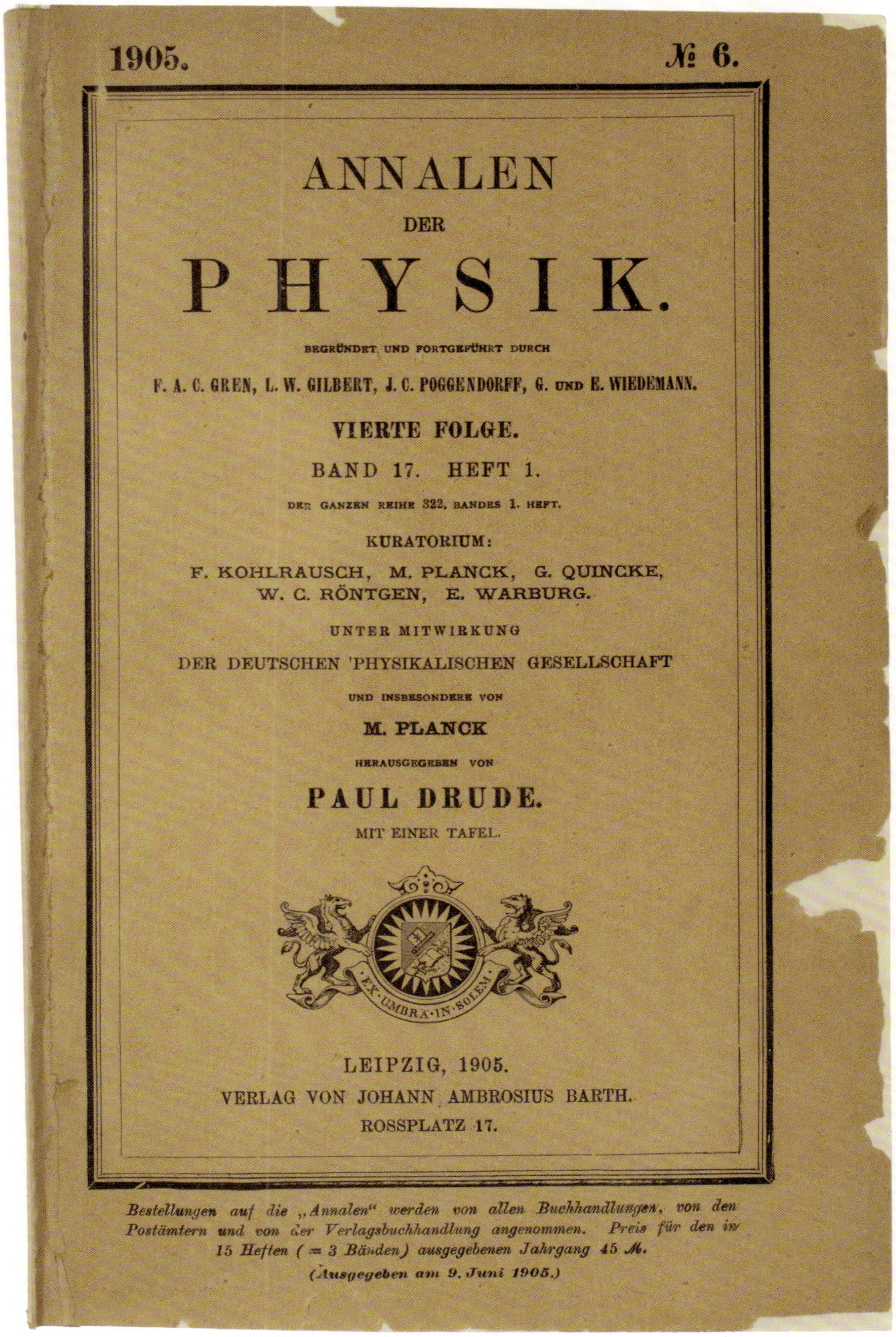

1905. № 6.

ANNALEN

DER

PHYSIK.

BEGRÜNDET UND FORTGEFÜHRT DURCH

F. A. C. GREN, L. W. GILBERT, J. C. POGGENDORFF, G. UND E. WIEDEMANN.

VIERTE FOLGE.

BAND 17. HEFT 1.

DER GANZEN REIHE 322. BANDES 1. HEFT.

KURATORIUM:

F. KOHLRAUSCH, M. PLANCK, G. QUINCKE,
W. C. RÖNTGEN, E. WARBURG.

UNTER MITWIRKUNG

DER DEUTSCHEN PHYSIKALISCHEN GESELLSCHAFT

UND INSBESONDERE VON

M. PLANCK

HERAUSGEGEBEN VON

PAUL DRUDE.

MIT EINER TAFEL.

EX·UMBRA·IN·SOLEM

LEIPZIG, 1905.
VERLAG VON JOHANN AMBROSIUS BARTH.
ROSSPLATZ 17.

*Bestellungen auf die „Annalen“ werden von allen Buchhandlungen, von den Postämtern und von der Verlagsbuchhandlung angenommen. Preis für den in 15 Heften (= 3 Bänden) ausgegebenen Jahrgang 45 ℳ.*

*(Ausgegeben am 9. Juni 1905.)*

*Pages 19–21:* VARIOUS AUTHORS,
Annalen der Physik (Annals of Physics), *1905.*

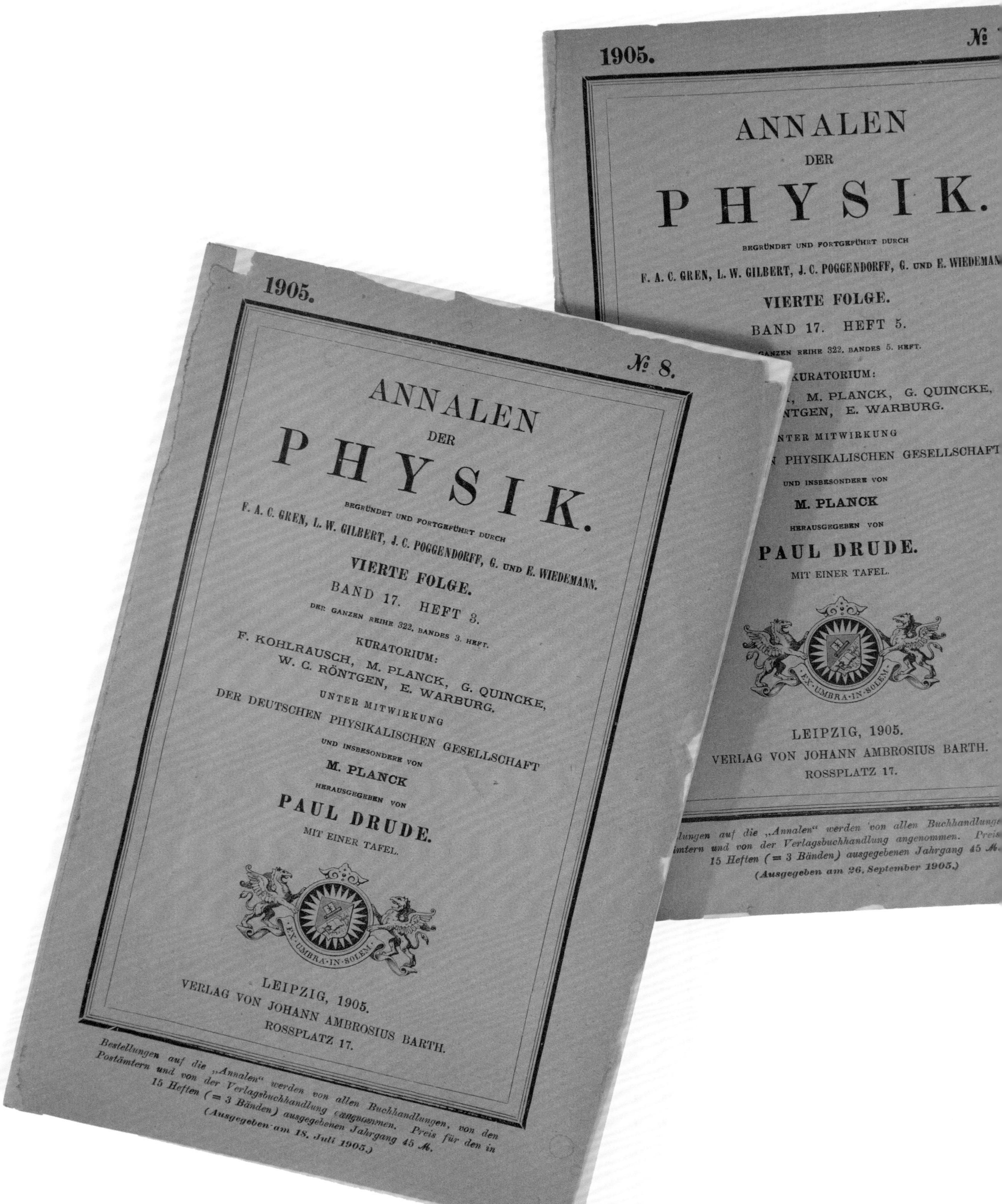
1905.
№ 8.
ANNALEN
DER
PHYSIK.
BEGRÜNDET UND FORTGEFÜHRT DURCH
F. A. C. GREN, L. W. GILBERT, J. C. POGGENDORFF, G. UND E. WIEDEMANN.
VIERTE FOLGE.
BAND 17. HEFT 3.
DER GANZEN REIHE 322. BANDES 3. HEFT.
KURATORIUM:
F. KOHLRAUSCH, M. PLANCK, G. QUINCKE,
W. C. RÖNTGEN, E. WARBURG.
UNTER MITWIRKUNG
DER DEUTSCHEN PHYSIKALISCHEN GESELLSCHAFT
UND INSBESONDERE VON
M. PLANCK
HERAUSGEGEBEN VON
PAUL DRUDE.
MIT EINER TAFEL.
EX·UMBRA·IN·SOLEM
LEIPZIG, 1905.
VERLAG VON JOHANN AMBROSIUS BARTH.
ROSSPLATZ 17.
Bestellungen auf die „Annalen" werden von allen Buchhandlungen, von den Postämtern und von der Verlagsbuchhandlung angenommen. Preis für den in 15 Heften (= 3 Bänden) ausgegebenen Jahrgang 45 ℳ.
(Ausgegeben am 18. Juli 1905.)
1905.
ANNALEN
DER
PHYSIK.
BEGRÜNDET UND FORTGEFÜHRT DURCH
F. A. C. GREN, L. W. GILBERT, J. C. POGGENDORFF, G. UND E. WIEDEMAN
VIERTE FOLGE.
BAND 17. HEFT 5.
GANZEN REIHE 322. BANDES 5. HEFT.
KURATORIUM:
M. PLANCK, G. QUINCKE,
NTGEN, E. WARBURG.
NTER MITWIRKUNG
N PHYSIKALISCHEN GESELLSCHAFT
UND INSBESONDERE VON
M. PLANCK
HERAUSGEGEBEN VON
PAUL DRUDE.
MIT EINER TAFEL.
EX·UMBRA·IN·SOLEM
LEIPZIG, 1905.
VERLAG VON JOHANN AMBROSIUS BARTH.
ROSSPLATZ 17.
llungen auf die „Annalen" werden von allen Buchhandlunge
ämtern und von der Verlagsbuchhandlung angenommen. Preis
15 Heften (= 3 Bänden) ausgegebenen Jahrgang 45 ℳ.
(Ausgegeben am 26. September 1905.)

*If a person could run after a light wave with the same speed as light, you would have a wave arrangement which could be completely independent of time. Of course, such a thing is impossible.”*

—ALBERT EINSTEIN, *1949*

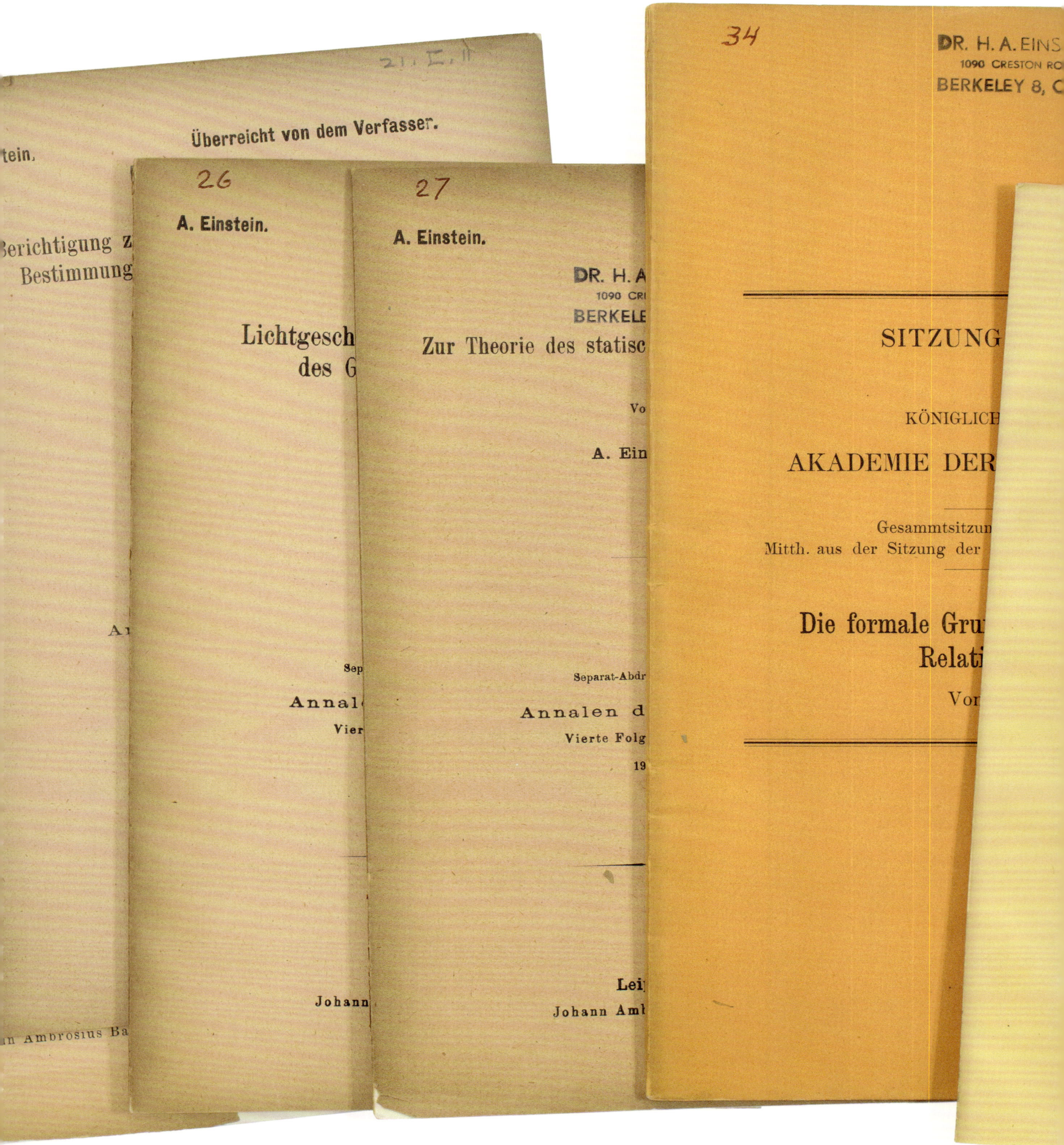

21. I. 11
Überreicht von dem Verfasser.
tein.
Berichtigung z
Bestimmung
An Ambrosius Ba
26
A. Einstein.
Lichtgesch
des G
Sep
Annal
Vier
Johann
27
A. Einstein.
DR. H. A
1090 CR
BERKELE
Zur Theorie des statisc
Vo
A. Ein
Separat-Abdr
Annalen d
Vierte Folg
19
Lei
Johann Aml
34
DR. H. A. EINS
1090 CRESTON RO
BERKELEY 8, C
SITZUNG
KÖNIGLICH
AKADEMIE DER
Gesammtsitzun
Mitth. aus der Sitzung der
Die formale Gru
Relati
Vor

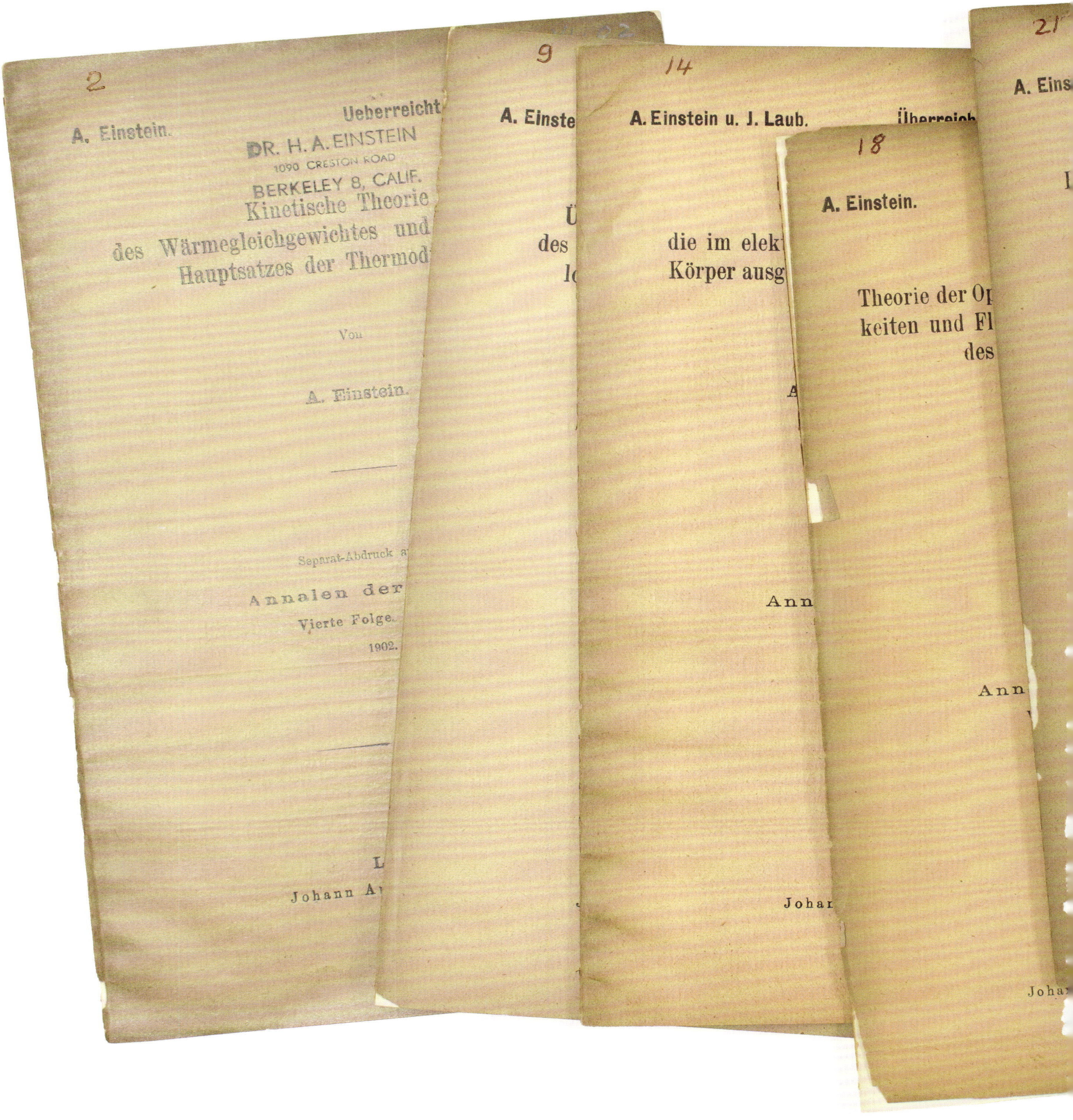
2
A. Einstein.
Ueberreicht
DR. H. A. EINSTEIN
1090 CRESTON ROAD
BERKELEY 8, CALIF.
Kinetische Theorie
des Wärmegleichgewichtes und
Hauptsatzes der Thermod
Von
A. Einstein.
Separat-Abdruck a
Annalen der
Vierte Folge.
1902.
Johann A
9
A. Einste
des
14
A. Einstein u. J. Laub.
die im elek
Körper ausg
Ann
18
A. Einstein.
Theorie der Op
keiten und Fl
des
Johan
21
A. Eins
Ann
Joha

## *EINSTEIN'S EQUIVALENCE PRINCIPLE*

After conceiving the special theory of relativity—limited to bodies in uniform motion—Einstein sought to generalize the theory to include gravitational motion. He called the principle that enabled him to do this the "equivalence principle." First described by Einstein in this 1907 paper, the equivalence principle asserted that the effects of gravity and acceleration are the same.

As he often did, Einstein proceeded with the aid of thought experiments. He imagined a man falling off a roof. While in free fall, the man experiences weightlessness, feeling no gravitational force acting on him. This "happiest thought" of Einstein's life led him to imagine other scenarios, such as a man enclosed in an elevator in space and unable to refer to any distant frame of reference. Einstein's thought experiments showed that locally (within the elevator), gravity and acceleration are indistinguishable by any observation or experiment the man can perform.

Einstein's thought experiments also revealed that an accelerating frame of reference would bend light; therefore, gravity will bend light according to the equivalence principle.

Years later, in his general theory of relativity, Einstein was able to quantify the degree to which massive objects bend space and time around them. He even proposed an experiment for astronomers to test his new theory. During a solar eclipse, it would be possible to measure the deflection in the apparent positions of stars visible in the darkened sky as their light passed nearby the sun. In 1919, Arthur Eddington carried out this test with his solar eclipse expedition, providing the first experimental proof of Einstein's revolutionary theory of gravity.

Jahrbuch der Radioaktivität und Elektronik
(Almanac of Radioactivity and Electronics), *1907.*
*Pages 24–27:* ALBERT EINSTEIN, *Offprints*
*from the library of Hans Albert Einstein.*

*"Einstein had already
published so many masterpieces,
none had actually been put to the test
and his theories were looked
on rather as tours de force than as
definitive additions to knowledge.
But his pre-eminence among the twelve
greatest theoretical physicists
of the day was clear
to any unprejudiced observer."*

—FREDERICK LINDEMANN

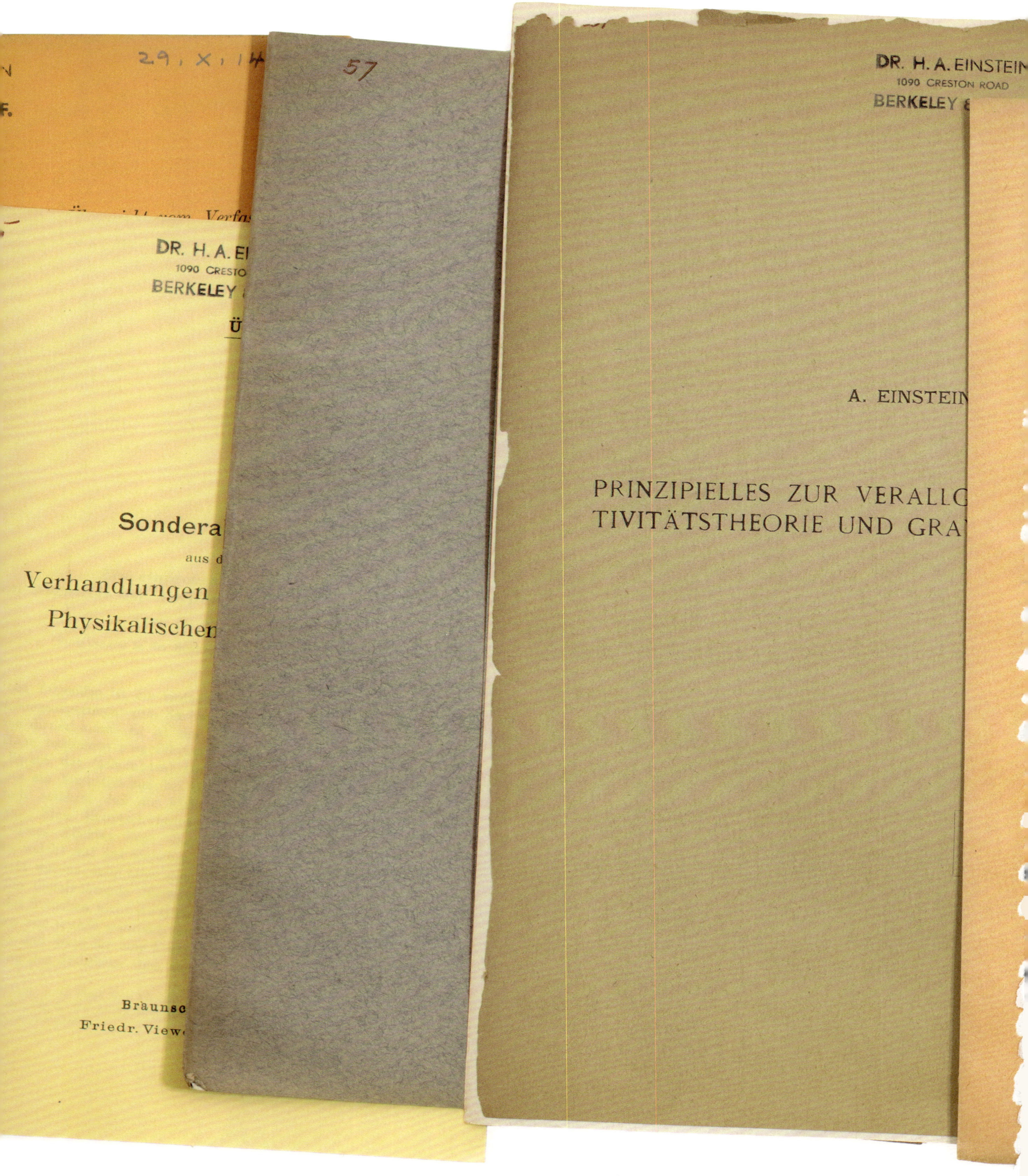
29. X. 14
57
DR. H. A. EINSTEIN
1090 CRESTON ROAD
BERKELEY
A. EINSTEIN
PRINZIPIELLES ZUR VERALLG
TIVITÄTSTHEORIE UND GRA
DR. H. A. EI
1090 CRESTO
BERKELEY
Sondera
Verhandlungen
Physikalischen
Braunsc
Friedr. View

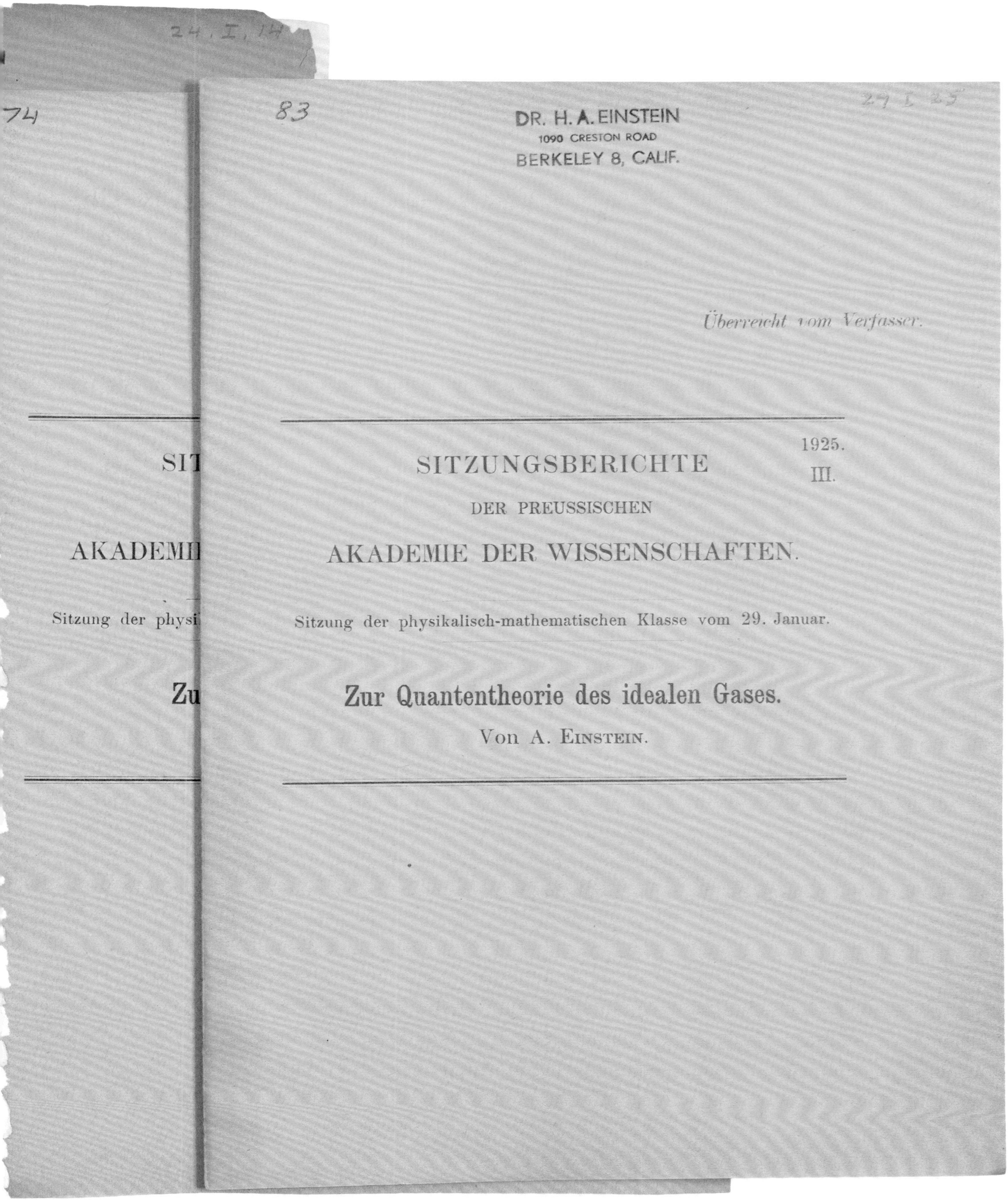

24. I. 14

74

SIT

AKADEMI

Sitzung der physi

Zu

83

DR. H. A. EINSTEIN
1090 CRESTON ROAD
BERKELEY 8, CALIF.

27 I 25

*Überreicht vom Verfasser.*

1925.
III.

SITZUNGSBERICHTE

DER PREUSSISCHEN

AKADEMIE DER WISSENSCHAFTEN.

Sitzung der physikalisch-mathematischen Klasse vom 29. Januar.

**Zur Quantentheorie des idealen Gases.**

Von A. Einstein.

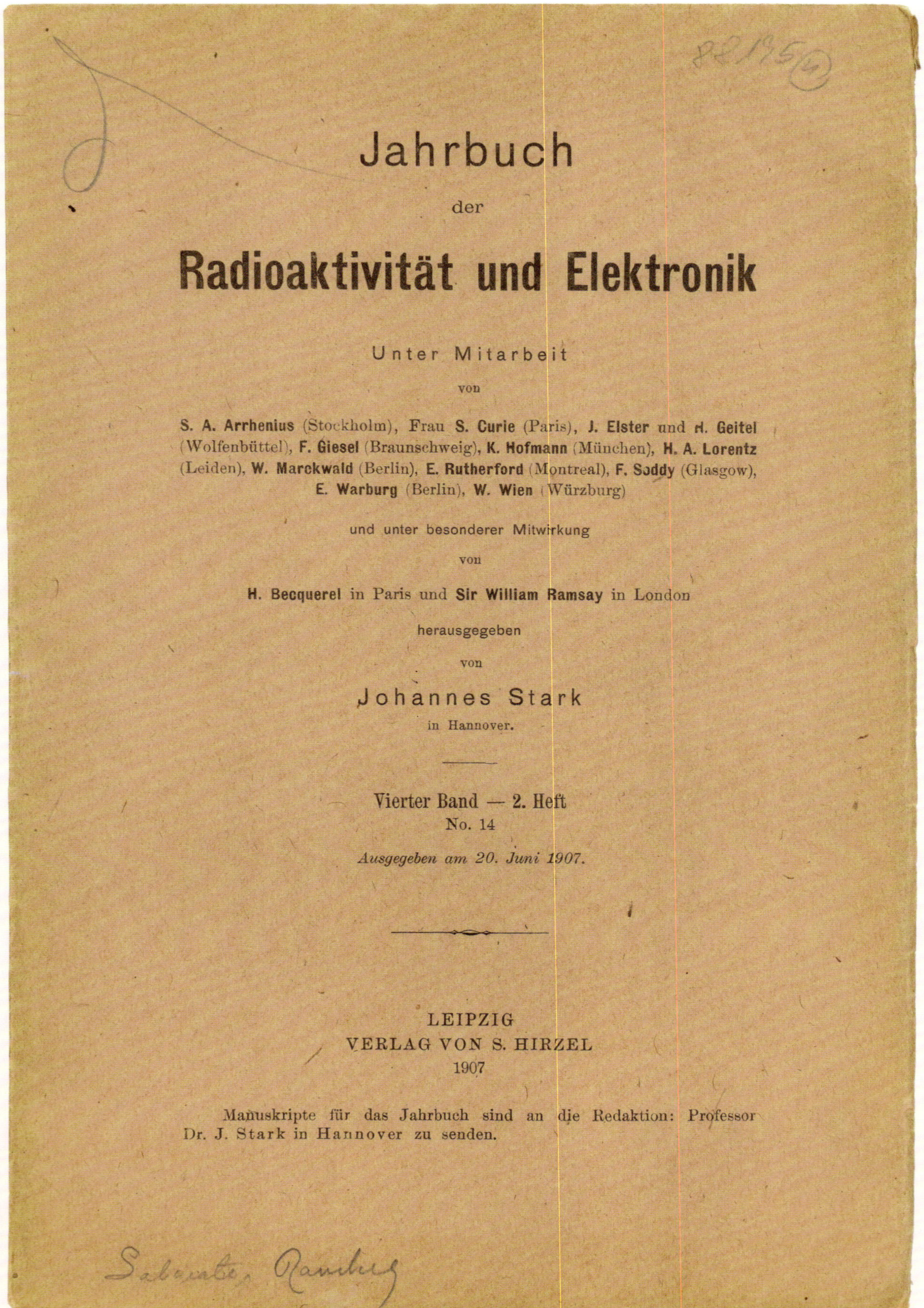

# Jahrbuch

der

# Radioaktivität und Elektronik

Unter Mitarbeit

von

**S. A. Arrhenius** (Stockholm), Frau **S. Curie** (Paris), **J. Elster** und **H. Geitel** (Wolfenbüttel), **F. Giesel** (Braunschweig), **K. Hofmann** (München), **H. A. Lorentz** (Leiden), **W. Marckwald** (Berlin), **E. Rutherford** (Montreal), **F. Soddy** (Glasgow), **E. Warburg** (Berlin), **W. Wien** (Würzburg)

und unter besonderer Mitwirkung

von

**H. Becquerel** in Paris und **Sir William Ramsay** in London

herausgegeben

von

## Johannes Stark

in Hannover.

Vierter Band — 2. Heft
No. 14

*Ausgegeben am 20. Juni 1907.*

LEIPZIG
VERLAG VON S. HIRZEL
1907

Manuskripte für das Jahrbuch sind an die Redaktion: Professor Dr. J. Stark in Hannover zu senden.

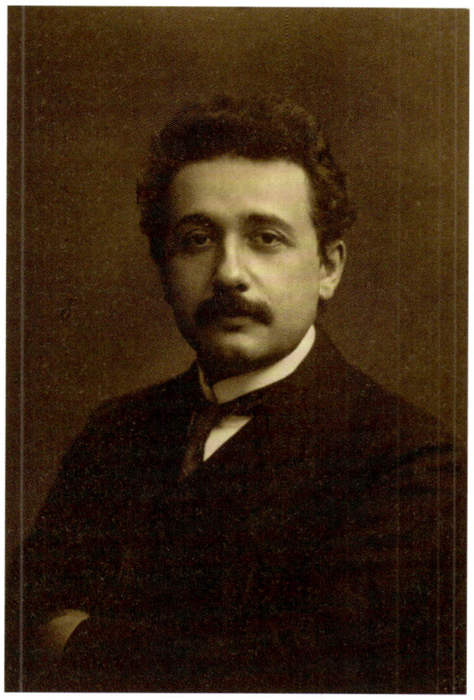

### *EINSTEIN ON THE INTERNATIONAL STAGE*

The First Solvay Conference was held in Brussels in 1911. It was attended by the world's leading physicists, most of whom were or would become Nobel laureates. This formal portrait of Einstein by photographer J. F. Langhans is included in an album with the other conference attendees.

Among the most renowned scientists of the day—including Ernest Rutherford, Marie Curie, and Max Planck—Einstein made a lasting impression. At age thirty-two, he was the second-youngest participant in the conference. The youngest was British physicist Frederick Lindemann.

According to Lindemann, although "Einstein had already published so many masterpieces, none had actually been put to the test and his theories were looked on rather as tours de force than as definitive additions to knowledge. But his pre-eminence among the twelve greatest theoretical physicists of the day was clear to any unprejudiced observer."

PHOTOGRAPH BY J. F. LANGHANS, *1911.*

## *THE MOST VALUABLE FIND*

On November 18, 1915, Einstein gave his third of four weekly lectures on general relativity to the Prussian Academy. The Academy printed his address, shown here, entitled (translated from the German) *Explanation of the Perihelion Movement of Mercury from the General Theory of Relativity.*

The perihelion is the nearest point to the Sun a planet reaches in its orbit. Astronomers had been aware for a long time of an "anomaly" in Mercury's perihelion, which shifts slightly with each revolution around the sun. The cause was a mystery. Even considering the effects of the other planets on Mercury's path through space, Newton's law of gravity could not explain this phenomonon. Newton's equations predicted an advance of only half the amount observed by astronomers. Einstein's equations, which showed how mass bends spacetime, yielded predictions that matched astronomers' observations.

Einstein said that when he saw his calculations accurately predicted the astronomical observations of Mercury's orbits, he experienced heart palpitations. He wrote to the physicist Arnold Sommerfeld: "It is the most valuable find that I have made in my life."

VARIOUS AUTHORS,
Sitzungsberichte der Königlich Preussischen Akademie der Wissenschaften zu Berlin (Reports from the Prussian Academy of Sciences in Berlin), *1915.*

1915 XLVII

# SITZUNGSBERICHTE

DER

KÖNIGLICH PREUSSISCHEN

# AKADEMIE DER WISSENSCHAFTEN



BERLIN 1915

VERLAG DER KÖNIGLICHEN AKADEMIE DER WISSENSCHAFTEN

IN KOMMISSION BEI GEORG REIMER

### *HIS MOST FAMOUS EQUATION*

In his 1905 paper entitled *Does the Inertia of a Body Depend on Its Energy Content?*, Einstein wrote (translated from German): "If a body gives off the energy L in the form of radiation, its mass diminishes by $L/c^2$ . . . the mass of a body is a measure of its energy-content." This statement, later represented by the elegant equation $E=mc^2$, would transform the world.

Since his teenage years, Einstein had puzzled over what a light beam would look like if he could catch up and travel alongside it. After a decade of pondering, Einstein postulated that the speed of light was the same for all observers independent of their frame of reference. In other words, the speed of light in a vacuum (c)—not the passage of time—was invariant throughout the universe. Einstein's brilliant insight led him to realize not only that time and space are relative but also that mass and energy are inseparable.

Before Einstein, mass and energy were thought to be distinct phenomena. Einstein realized they are different properties of the same thing. He discovered purely by thought that the mass of a body at rest is a measure of its energy content. The energy value can be computed with a single constant: the speed of light. Because of the enormous value of the speed of light when squared, a tiny amount of mass was shown to contain a huge amount of energy.

Einstein's most simple and famous equation led to an understanding of radioactive decay and the energy source of stars. It is the basis of harnessing nuclear energy for power and electricity.

Tragically, $E=mc^2$ also provided the theoretical basis of atomic bomb research. Einstein had given support to the research with his famous wartime letter to President Roosevelt. It was a decision he soon came to regret. When Einstein heard that the bomb had been used in Japan, he lamented, "Woe is me." Just months before his death he said, "I made one mistake in my life when I signed that letter to President Roosevelt advocating that the atomic bomb should be built. But, perhaps I can be forgiven for that because we all felt that there was a high probability that the Germans were working on this problem and they might succeed and use the atomic bomb to become the master race."

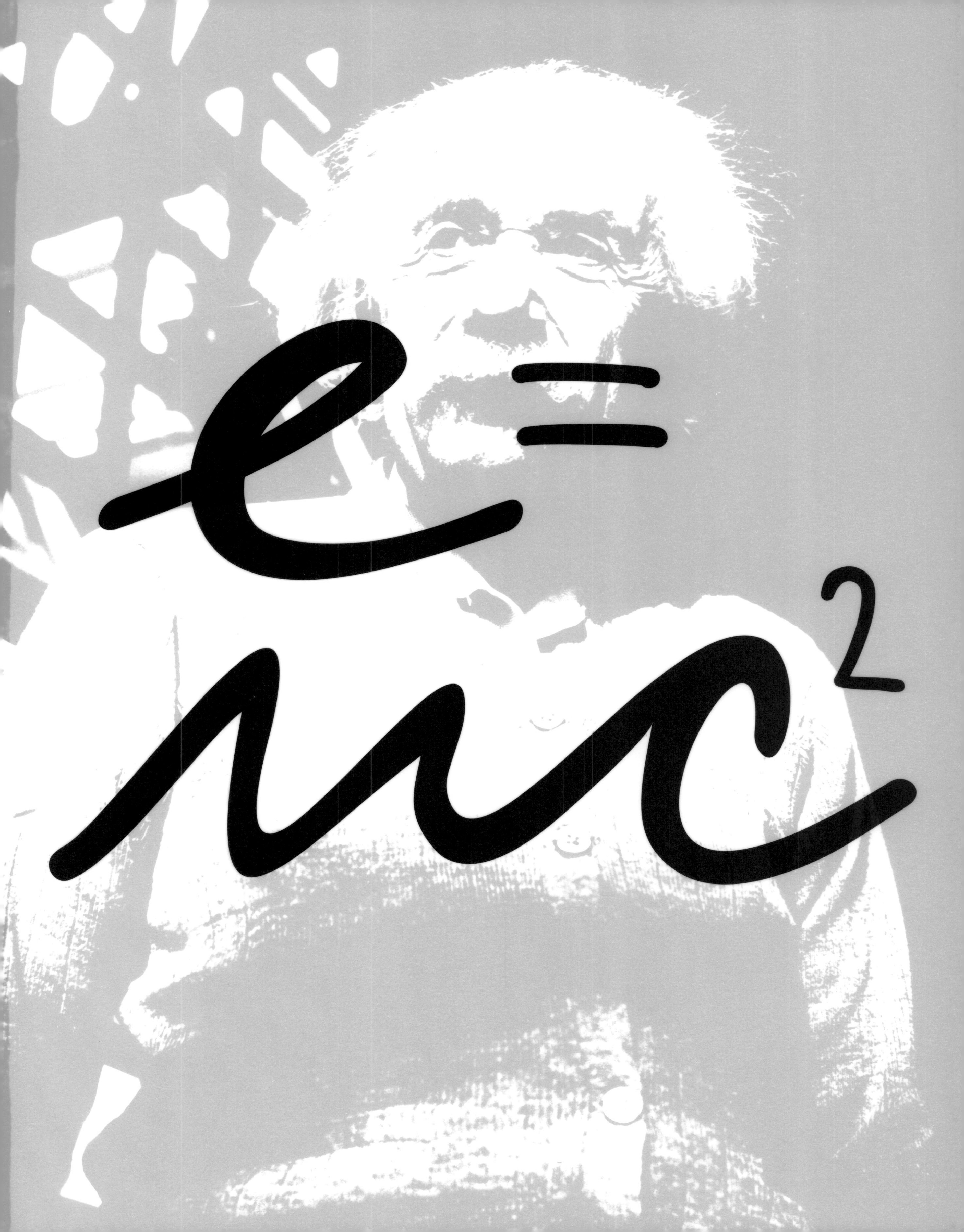
$e=mc^2$

## *EINSTEIN'S CROWNING ACHIEVEMENT*

"One of the greatest achievements in the history of human thought. It is not the discovery of an outlying island but of a whole continent of new scientific ideas."
—J. J. Thomson, Nobel laureate and president of the Royal Society

After a decade of strenuous effort, Einstein succeeded in generalizing his theory of relativity from the special case of bodies in uniform motion to describe all motion, including bodies in accelerated motion. Having completed his general theory of relativity by the end of 1915, Einstein published the finalized theory in the April 1916 issue of *Annalen der Physik* and in a separate printing, shown here.

According to general relativity, gravity is not Newton's mysterious force acting instantaneously between distant bodies; it's the acceleration of mass resulting from how bodies travel naturally through the curvature in spacetime they create.

Einstein's new theory was a brilliant leap of imagination, developed through a secure grounding in the mathematics of tensor calculus and Riemannian geometry. The "field equations" Einstein ultimately created for his general theory described the contours of spacetime and the behavior of matter within it. According to the theory of general relativity, the American theoretical physicist John Wheeler observed, "Mass tells space-time how to curve, and space-time tells mass how to move."

The implications of his revolutionary theory were astounding. The principles of general relativity extended to the behavior of the entire universe. It helped explain the motion of planets and the formation of black holes. It provided evidence for the Big Bang, predicted the existence of gravitational waves and gravitational lensing, and has been essential in practical applications, such as modern GPS systems.

Einstein was aware of the profound power of his new theory. In a letter to his good friend Michele Besso in late 1915, he triumphantly announced that his "boldest dreams have now been fulfilled."

*Pages 35–39:* ALBERT EINSTEIN,
Die Grundlage der allgemeinen Relativitätstheorie
(The General Theory of Relativity), *1916.*

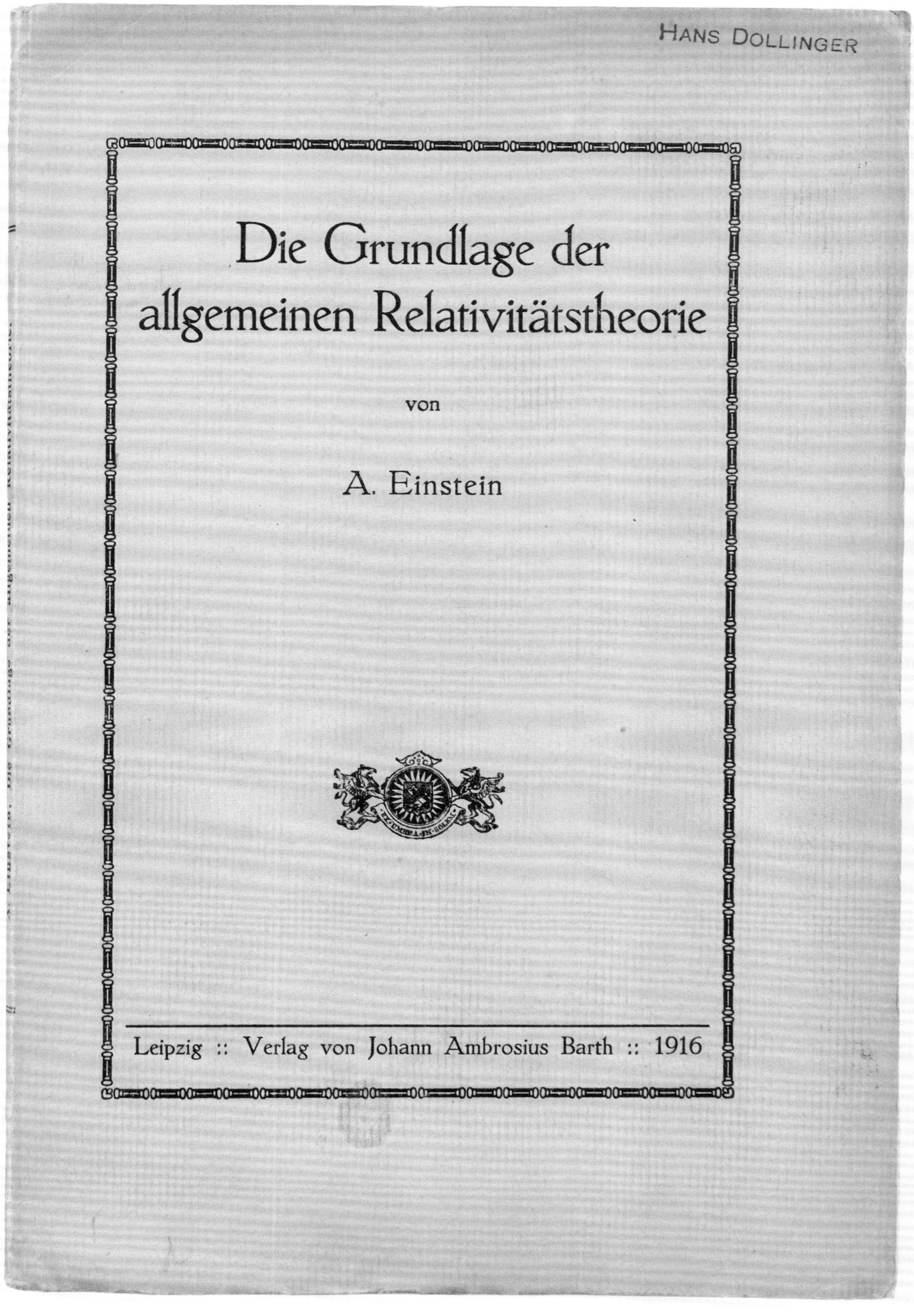

Hans Dollinger

Die Grundlage der allgemeinen Relativitätstheorie

von

A. Einstein

Leipzig :: Verlag von Johann Ambrosius Barth :: 1916

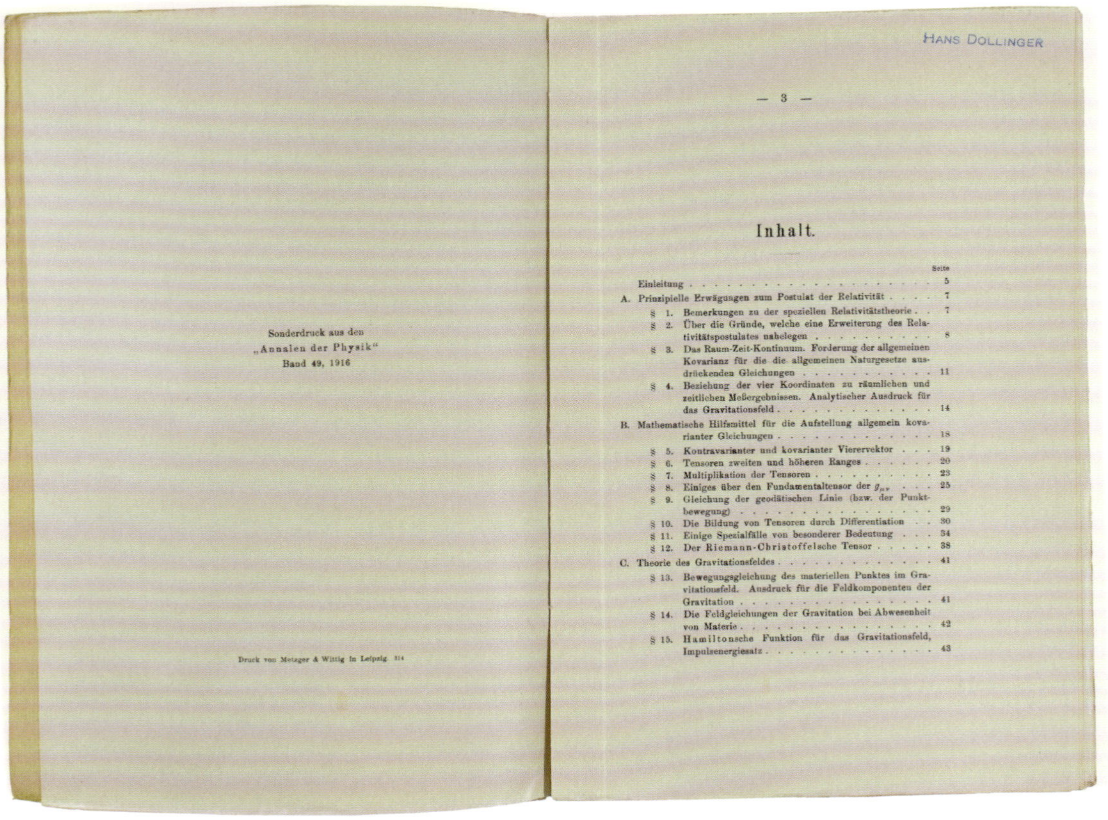

Sonderdruck aus den
„Annalen der Physik"
Band 49, 1916

Druck von Metzger & Wittig in Leipzig 314

HANS DOLLINGER

— 3 —

Inhalt.



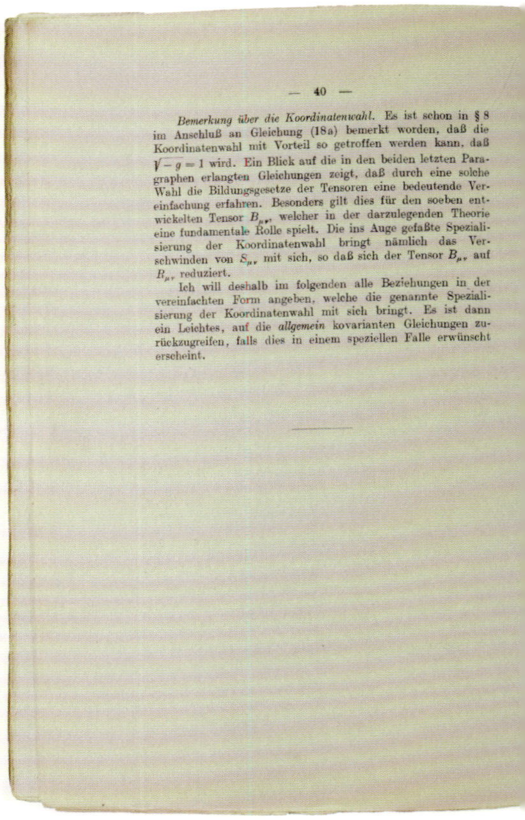

— 40 —

*Bemerkung über die Koordinatenwahl.* Es ist schon in § 8 im Anschluß an Gleichung (18a) bemerkt worden, daß die Koordinatenwahl mit Vorteil so getroffen werden kann, daß $\sqrt{-g} = 1$ wird. Ein Blick auf die in den beiden letzten Paragraphen erlangten Gleichungen zeigt, daß durch eine solche Wahl die Bildungsgesetze der Tensoren eine bedeutende Vereinfachung erfahren. Besonders gilt dies für den soeben entwickelten Tensor $B_{\mu\nu}$, welcher in der darzulegenden Theorie eine fundamentale Rolle spielt. Die ins Auge gefaßte Spezialisierung der Koordinatenwahl bringt nämlich das Verschwinden von $S_{\mu\nu}$ mit sich, so daß sich der Tensor $B_{\mu\nu}$ auf $R_{\mu\nu}$ reduziert.

Ich will deshalb im folgenden alle Beziehungen in der vereinfachten Form angeben, welche die genannte Spezialisierung der Koordinatenwahl mit sich bringt. Es ist dann ein Leichtes, auf die *allgemein* kovarianten Gleichungen zurückzugreifen, falls dies in einem speziellen Falle erwünscht erscheint.

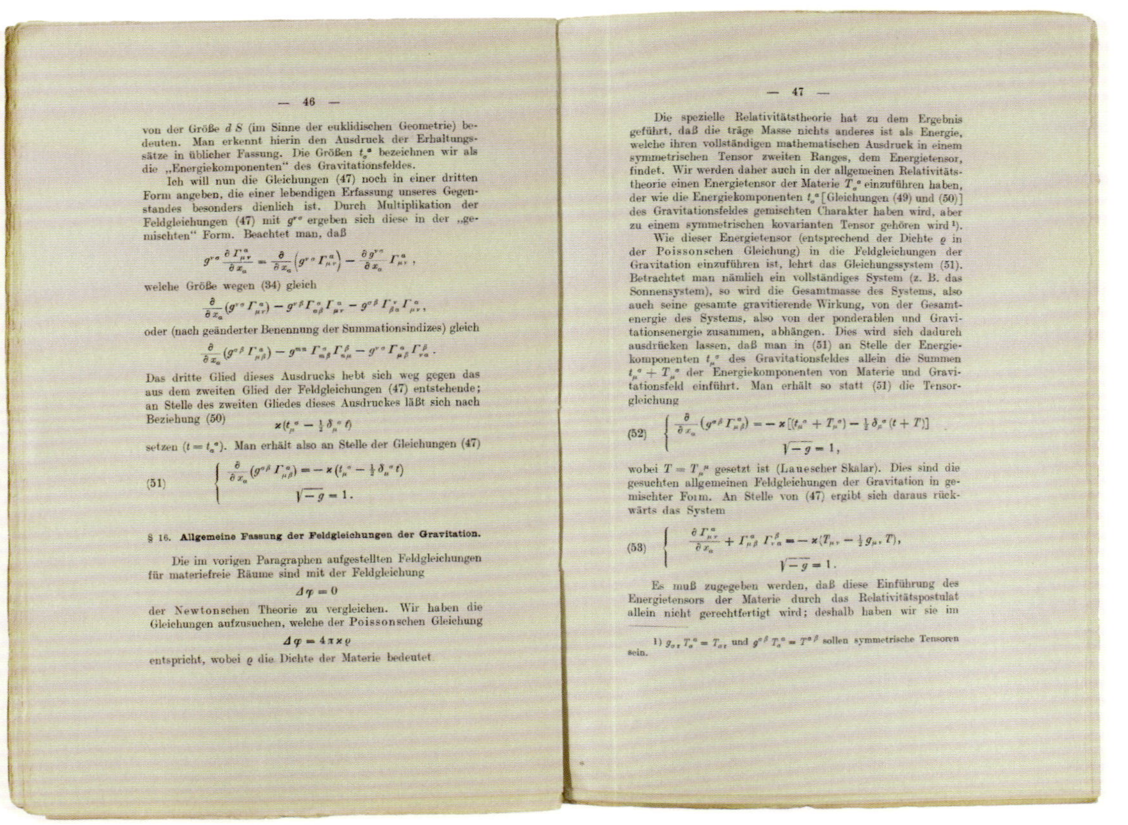

— 46 —

von der Größe $dS$ (im Sinne der euklidischen Geometrie) bedeuten. Man erkennt hierin den Ausdruck der Erhaltungssätze in üblicher Fassung. Die Größen $t_\sigma^\alpha$ bezeichnen wir als die „Energiekomponenten" des Gravitationsfeldes.

Ich will nun die Gleichungen (47) noch in einer dritten Form angeben, die einer lebendigen Erfassung unseres Gegenstandes besonders dienlich ist. Durch Multiplikation der Feldgleichungen (47) mit $g^{\nu\sigma}$ ergeben sich diese in der „gemischten" Form. Beachtet man, daß

$$g^{\nu\sigma}\frac{\partial \Gamma^\alpha_{\mu\nu}}{\partial x_\alpha} = \frac{\partial}{\partial x_\alpha}\left(g^{\nu\sigma}\Gamma^\alpha_{\mu\nu}\right) - \frac{\partial g^{\nu\sigma}}{\partial x_\alpha}\Gamma^\alpha_{\mu\nu},$$

welche Größe wegen (34) gleich

$$\frac{\partial}{\partial x_\alpha}\left(g^{\nu\sigma}\Gamma^\alpha_{\mu\nu}\right) - g^{\nu\beta}\Gamma^\sigma_{\alpha\beta}\Gamma^\alpha_{\mu\nu} - g^{\sigma\beta}\Gamma^\nu_{\beta\alpha}\Gamma^\alpha_{\mu\nu},$$

oder (nach geänderter Benennung der Summationsindizes) gleich

$$\frac{\partial}{\partial x_\alpha}\left(g^{\sigma\beta}\Gamma^\alpha_{\mu\beta}\right) - g^{mn}\Gamma^\sigma_{m\beta}\Gamma^\beta_{n\mu} - g^{\nu\sigma}\Gamma^\alpha_{\mu\beta}\Gamma^\beta_{\nu\alpha}.$$

Das dritte Glied dieses Ausdrucks hebt sich weg gegen das aus dem zweiten Glied der Feldgleichungen (47) entstehende; an Stelle des zweiten Gliedes dieses Ausdruckes läßt sich nach Beziehung (50)

$$\varkappa\left(t_\mu^\sigma - \tfrac{1}{2}\delta_\mu^\sigma t\right)$$

setzen ($t = t_\alpha^\alpha$). Man erhält also an Stelle der Gleichungen (47)

$$(51)\qquad \begin{cases} \dfrac{\partial}{\partial x_\alpha}\left(g^{\sigma\beta}\Gamma^\alpha_{\mu\beta}\right) = -\varkappa\left(t_\mu^\sigma - \tfrac{1}{2}\delta_\mu^\sigma t\right) \\ \sqrt{-g} = 1. \end{cases}$$

§ 16. **Allgemeine Fassung der Feldgleichungen der Gravitation.**

Die im vorigen Paragraphen aufgestellten Feldgleichungen für materiefreie Räume sind mit der Feldgleichung

$$\Delta\varphi = 0$$

der Newtonschen Theorie zu vergleichen. Wir haben die Gleichungen aufzusuchen, welche der Poissonschen Gleichung

$$\Delta\varphi = 4\pi\varkappa\varrho$$

entspricht, wobei $\varrho$ die Dichte der Materie bedeutet.

— 47 —

Die spezielle Relativitätstheorie hat zu dem Ergebnis geführt, daß die träge Masse nichts anderes ist als Energie, welche ihren vollständigen mathematischen Ausdruck in einem symmetrischen Tensor zweiten Ranges, dem Energietensor, findet. Wir werden daher auch in der allgemeinen Relativitätstheorie einen Energietensor der Materie $T_\sigma^\alpha$ einzuführen haben, der wie die Energiekomponenten $t_\sigma^\alpha$ [Gleichungen (49) und (50)] des Gravitationsfeldes gemischten Charakter haben wird, aber zu einem symmetrischen kovarianten Tensor gehören wird[1]).

Wie dieser Energietensor (entsprechend der Dichte $\varrho$ in der Poissonschen Gleichung) in die Feldgleichungen der Gravitation einzuführen ist, lehrt das Gleichungssystem (51). Betrachtet man nämlich ein vollständiges System (z. B. das Sonnensystem), so wird die Gesamtmasse des Systems, also auch seine gesamte gravitierende Wirkung, von der Gesamtenergie des Systems, also von der ponderablen und Gravitationsenergie zusammen, abhängen. Dies wird sich dadurch ausdrücken lassen, daß man in (51) an Stelle der Energiekomponenten $t_\mu^\sigma$ des Gravitationsfeldes allein die Summen $t_\mu^\sigma + T_\mu^\sigma$ der Energiekomponenten von Materie und Gravitationsfeld einführt. Man erhält so statt (51) die Tensorgleichung

$$(52)\qquad \begin{cases} \dfrac{\partial}{\partial x_\alpha}\left(g^{\sigma\beta}\Gamma^\alpha_{\mu\beta}\right) = -\varkappa\left[\left(t_\mu^\sigma + T_\mu^\sigma\right) - \tfrac{1}{2}\delta_\mu^\sigma(t+T)\right] \\ \sqrt{-g} = 1, \end{cases}$$

wobei $T = T_\mu^\mu$ gesetzt ist (Lauescher Skalar). Dies sind die gesuchten allgemeinen Feldgleichungen der Gravitation in gemischter Form. An Stelle von (47) ergibt sich daraus rückwärts das System

$$(53)\qquad \begin{cases} \dfrac{\partial \Gamma^\alpha_{\mu\nu}}{\partial x_\alpha} + \Gamma^\alpha_{\mu\beta}\Gamma^\beta_{\nu\alpha} = -\varkappa\left(T_{\mu\nu} - \tfrac{1}{2}g_{\mu\nu}T\right), \\ \sqrt{-g} = 1. \end{cases}$$

Es muß zugegeben werden, daß diese Einführung des Energietensors der Materie durch das Relativitätspostulat allein nicht gerechtfertigt wird; deshalb haben wir sie im

1) $g_{\sigma\tau}T_\alpha^\sigma = T_{\alpha\tau}$ und $g^{\sigma\beta}T_\sigma^\alpha = T^{\alpha\beta}$ sollen symmetrische Tensoren sein.

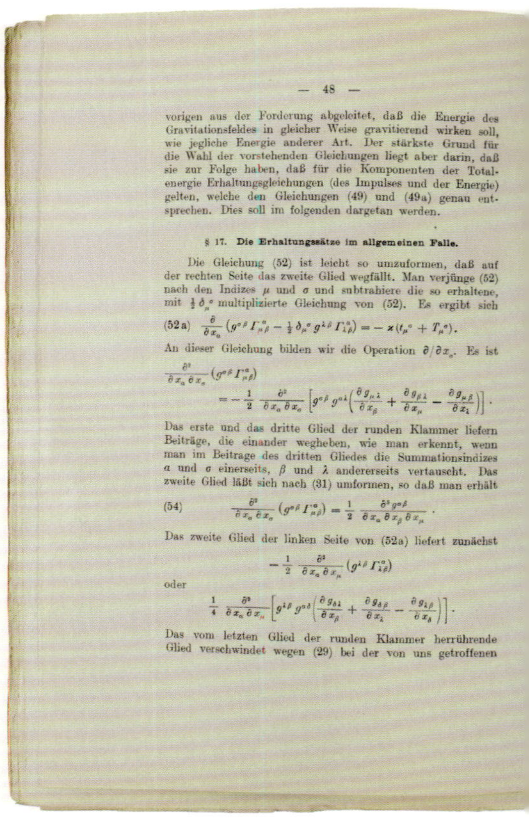

— 48 —

vorigen aus der Forderung abgeleitet, daß die Energie des Gravitationsfeldes in gleicher Weise gravitierend wirken soll, wie jegliche Energie anderer Art. Der stärkste Grund für die Wahl der vorstehenden Gleichungen liegt aber darin, daß sie zur Folge haben, daß für die Komponenten der Totalenergie Erhaltungsgleichungen (des Impulses und der Energie) gelten, welche den Gleichungen (49) und (49a) genau entsprechen. Dies soll im folgenden dargetan werden.

§ 17. **Die Erhaltungssätze im allgemeinen Falle.**

Die Gleichung (52) ist leicht so umzuformen, daß auf der rechten Seite das zweite Glied wegfällt. Man verjünge (52) nach den Indizes $\mu$ und $\sigma$ und subtrahiere die so erhaltene, mit $\frac{1}{2}\delta_\mu^\sigma$ multiplizierte Gleichung von (52). Es ergibt sich

$$(52\text{a})\qquad \frac{\partial}{\partial x_\alpha}\left(g^{\sigma\beta}\Gamma^\alpha_{\mu\beta} - \tfrac{1}{2}\delta_\mu^\sigma g^{\lambda\beta}\Gamma^\alpha_{\lambda\beta}\right) = -\varkappa\left(t_\mu^\sigma + T_\mu^\sigma\right).$$

An dieser Gleichung bilden wir die Operation $\partial/\partial x_\sigma$. Es ist

$$\frac{\partial^2}{\partial x_\alpha \partial x_\sigma}\left(g^{\sigma\beta}\Gamma^\alpha_{\mu\beta}\right) = -\frac{1}{2}\frac{\partial^2}{\partial x_\alpha \partial x_\sigma}\left[g^{\sigma\beta}g^{\alpha\lambda}\left(\frac{\partial g_{\mu\lambda}}{\partial x_\beta} + \frac{\partial g_{\beta\lambda}}{\partial x_\mu} - \frac{\partial g_{\mu\beta}}{\partial x_\lambda}\right)\right].$$

Das erste und das dritte Glied der runden Klammer liefern Beiträge, die einander wegheben, wie man erkennt, wenn man im Beitrage des dritten Gliedes die Summationsindizes $\alpha$ und $\sigma$ einerseits, $\beta$ und $\lambda$ andererseits vertauscht. Das zweite Glied läßt sich nach (31) umformen, so daß man erhält

$$(54)\qquad \frac{\partial^2}{\partial x_\alpha \partial x_\sigma}\left(g^{\sigma\beta}\Gamma^\alpha_{\mu\beta}\right) = \frac{1}{2}\frac{\partial^3 g^{\alpha\beta}}{\partial x_\alpha \partial x_\beta \partial x_\mu}.$$

Das zweite Glied der linken Seite von (52a) liefert zunächst

$$-\frac{1}{2}\frac{\partial^2}{\partial x_\alpha \partial x_\mu}\left(g^{\lambda\beta}\Gamma^\alpha_{\lambda\beta}\right)$$

oder

$$\frac{1}{4}\frac{\partial^2}{\partial x_\alpha \partial x_\mu}\left[g^{\lambda\beta}g^{\alpha\delta}\left(\frac{\partial g_{\delta\lambda}}{\partial x_\beta} + \frac{\partial g_{\delta\beta}}{\partial x_\lambda} - \frac{\partial g_{\lambda\beta}}{\partial x_\delta}\right)\right].$$

Das vom letzten Glied der runden Klammer herrührende Glied verschwindet wegen (29) bei der von uns getroffenen

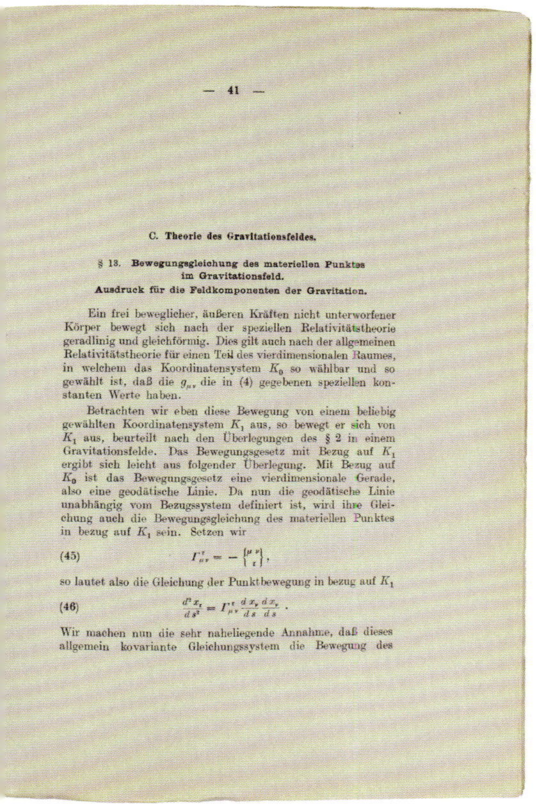

— 41 —

C. Theorie des Gravitationsfeldes.

§ 13. Bewegungsgleichung des materiellen Punktes im Gravitationsfeld. Ausdruck für die Feldkomponenten der Gravitation.

Ein frei beweglicher, äußeren Kräften nicht unterworfener Körper bewegt sich nach der speziellen Relativitätstheorie geradlinig und gleichförmig. Dies gilt auch nach der allgemeinen Relativitätstheorie für einen Teil des vierdimensionalen Raumes, in welchem das Koordinatensystem $K_0$ so wählbar und so gewählt ist, daß die $g_{\mu\nu}$ die in (4) gegebenen speziellen konstanten Werte haben.

Betrachten wir eben diese Bewegung von einem beliebig gewählten Koordinatensystem $K_1$ aus, so bewegt er sich von $K_1$ aus, beurteilt nach den Überlegungen des § 2 in einem Gravitationsfelde. Das Bewegungsgesetz mit Bezug auf $K_1$ ergibt sich leicht aus folgender Überlegung. Mit Bezug auf $K_0$ ist das Bewegungsgesetz eine vierdimensionale Gerade, also eine geodätische Linie. Da nun die geodätische Linie unabhängig vom Bezugssystem definiert ist, wird ihre Gleichung auch die Bewegungsgleichung des materiellen Punktes in bezug auf $K_1$ sein. Setzen wir

$$\Gamma^{\tau}_{\mu\nu} = -\begin{Bmatrix}\mu\nu\\ \tau\end{Bmatrix}, \tag{45}$$

so lautet also die Gleichung der Punktbewegung in bezug auf $K_1$

$$\frac{d^2 x_\tau}{ds^2} = \Gamma^{\tau}_{\mu\nu}\frac{dx_\mu}{ds}\frac{dx_\nu}{ds}. \tag{46}$$

Wir machen nun die sehr naheliegende Annahme, daß dieses allgemein kovariante Gleichungssystem die Bewegung des

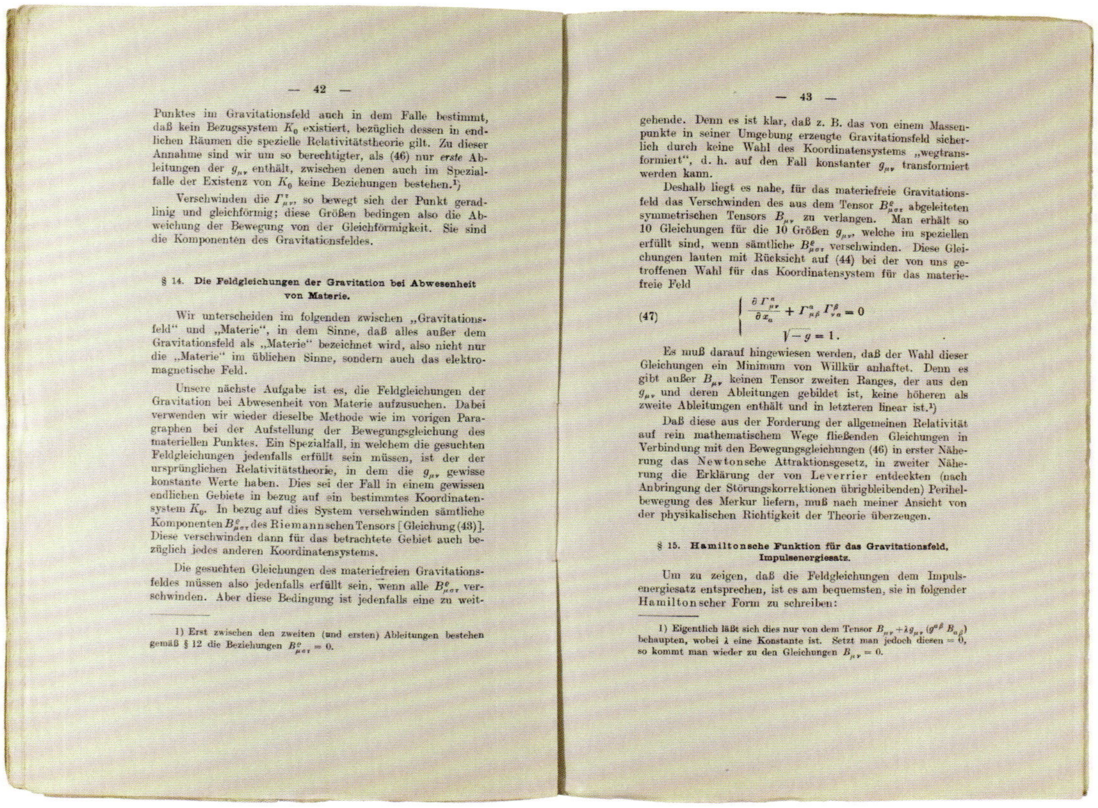

— 42 —

Punktes im Gravitationsfeld auch in dem Falle bestimmt, daß kein Bezugssystem $K_0$ existiert, bezüglich dessen in endlichen Räumen die spezielle Relativitätstheorie gilt. Zu dieser Annahme sind wir um so berechtigter, als (46) nur *erste* Ableitungen der $g_{\mu\nu}$ enthält, zwischen denen auch im Spezialfalle der Existenz von $K_0$ keine Beziehungen bestehen.[1]

Verschwinden die $\Gamma^{\tau}_{\mu\nu}$, so bewegt sich der Punkt geradlinig und gleichförmig; diese Größen bedingen also die Abweichung der Bewegung von der Gleichförmigkeit. Sie sind die Komponenten des Gravitationsfeldes.

§ 14. Die Feldgleichungen der Gravitation bei Abwesenheit von Materie.

Wir unterscheiden im folgenden zwischen „Gravitationsfeld" und „Materie", in dem Sinne, daß alles außer dem Gravitationsfeld als „Materie" bezeichnet wird, also nicht nur die „Materie" im üblichen Sinne, sondern auch das elektromagnetische Feld.

Unsere nächste Aufgabe ist es, die Feldgleichungen der Gravitation bei Abwesenheit von Materie aufzusuchen. Dabei verwenden wir wieder dieselbe Methode wie im vorigen Paragraphen bei der Aufstellung der Bewegungsgleichung des materiellen Punktes. Ein Spezialfall, in welchem die gesuchten Feldgleichungen jedenfalls erfüllt sein müssen, ist der der ursprünglichen Relativitätstheorie, in dem die $g_{\mu\nu}$ gewisse konstante Werte haben. Dies sei der Fall in einem gewissen endlichen Gebiete in bezug auf ein bestimmtes Koordinatensystem $K_0$. In bezug auf dies System verschwinden sämtliche Komponenten $B^{\varrho}_{\mu\sigma\tau}$ des Riemannschen Tensors [Gleichung (43)]. Diese verschwinden dann für das betrachtete Gebiet auch bezüglich jedes anderen Koordinatensystems.

Die gesuchten Gleichungen des materiefreien Gravitationsfeldes müssen also jedenfalls erfüllt sein, wenn alle $B^{\varrho}_{\mu\sigma\tau}$ verschwinden. Aber diese Bedingung ist jedenfalls eine zu weit-

1) Erst zwischen den zweiten (und ersten) Ableitungen bestehen gemäß § 12 die Beziehungen $B^{\varrho}_{\mu\sigma\tau} = 0$.

— 43 —

gehende. Denn es ist klar, daß z. B. das von einem Massenpunkte in seiner Umgebung erzeugte Gravitationsfeld sicherlich durch keine Wahl des Koordinatensystems „wegtransformiert", d. h. auf den Fall konstanter $g_{\mu\nu}$ transformiert werden kann.

Deshalb liegt es nahe, für das materiefreie Gravitationsfeld das Verschwinden des aus dem Tensor $B^{\varrho}_{\mu\sigma\tau}$ abgeleiteten symmetrischen Tensors $B_{\mu\nu}$ zu verlangen. Man erhält so 10 Gleichungen für die 10 Größen $g_{\mu\nu}$, welche im speziellen erfüllt sind, wenn sämtliche $B^{\varrho}_{\mu\sigma\tau}$ verschwinden. Diese Gleichungen lauten mit Rücksicht auf (44) bei der von uns getroffenen Wahl für das Koordinatensystem für das materiefreie Feld

$$\left\{\begin{aligned} &\frac{\partial \Gamma^{\alpha}_{\mu\nu}}{\partial x_\alpha} + \Gamma^{\alpha}_{\mu\beta}\Gamma^{\beta}_{\nu\alpha} = 0 \\ &\sqrt{-g} = 1. \end{aligned}\right. \tag{47}$$

Es muß darauf hingewiesen werden, daß der Wahl dieser Gleichungen ein Minimum von Willkür anhaftet. Denn es gibt außer $B_{\mu\nu}$ keinen Tensor zweiten Ranges, der aus den $g_{\mu\nu}$ und deren Ableitungen gebildet ist, keine höheren als zweite Ableitungen enthält und in letzteren linear ist.[1]

Daß diese aus der Forderung der allgemeinen Relativität auf rein mathematischem Wege fließenden Gleichungen in Verbindung mit den Bewegungsgleichungen (46) in erster Näherung das Newtonsche Attraktionsgesetz, in zweiter Näherung die Erklärung der von Leverrier entdeckten (nach Anbringung der Störungskorrektionen übrigbleibenden) Perihelbewegung des Merkur liefern, muß nach meiner Ansicht von der physikalischen Richtigkeit der Theorie überzeugen.

§ 15. Hamiltonsche Funktion für das Gravitationsfeld, Impulsenergiesatz.

Um zu zeigen, daß die Feldgleichungen dem Impulsenergiesatz entsprechen, ist es am bequemsten, sie in folgender Hamiltonscher Form zu schreiben:

1) Eigentlich läßt sich dies nur von dem Tensor $B_{\mu\nu} + \lambda g_{\mu\nu}(g^{\alpha\beta}B_{\alpha\beta})$ behaupten, wobei $\lambda$ eine Konstante ist. Setzt man jedoch diesen $= 0$, so kommt man wieder zu den Gleichungen $B_{\mu\nu} = 0$.

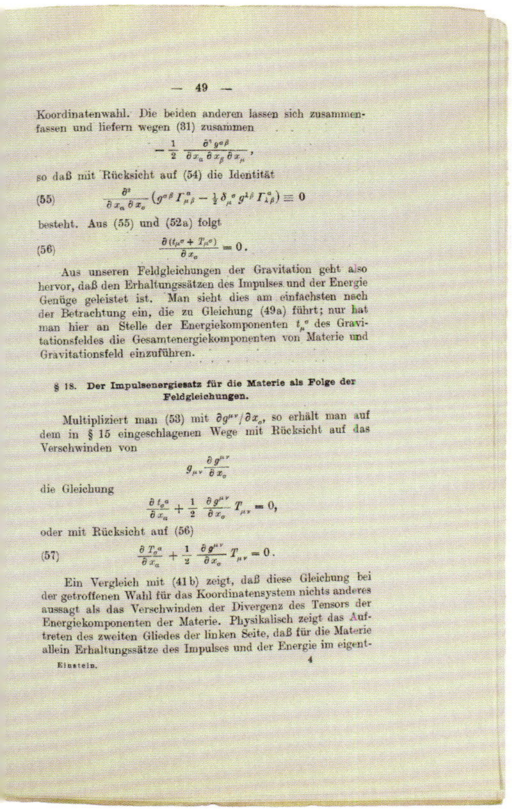

— 49 —

Koordinatenwahl. Die beiden anderen lassen sich zusammenfassen und liefern wegen (31) zusammen

$$-\frac{1}{2}\frac{\partial^3 g^{\alpha\beta}}{\partial x_\alpha \partial x_\beta \partial x_\mu},$$

so daß mit Rücksicht auf (54) die Identität

$$\frac{\partial^2}{\partial x_\alpha \partial x_\sigma}\left(g^{\sigma\beta}\Gamma^{\alpha}_{\mu\beta} - \tfrac{1}{2}\delta_\mu^\sigma g^{\lambda\beta}\Gamma^{\alpha}_{\lambda\beta}\right) \equiv 0 \tag{55}$$

besteht. Aus (55) und (52a) folgt

$$\frac{\partial (t_\mu^\sigma + T_\mu^\sigma)}{\partial x_\sigma} = 0. \tag{56}$$

Aus unseren Feldgleichungen der Gravitation geht also hervor, daß den Erhaltungssätzen des Impulses und der Energie Genüge geleistet ist. Man sieht dies am einfachsten nach der Betrachtung ein, die zu Gleichung (49a) führt; nur hat man hier an Stelle der Energiekomponenten $t_\mu^\sigma$ des Gravitationsfeldes die Gesamtenergiekomponenten von Materie und Gravitationsfeld einzuführen.

§ 18. Der Impulsenergiesatz für die Materie als Folge der Feldgleichungen.

Multipliziert man (53) mit $\partial g^{\mu\nu}/\partial x_\sigma$, so erhält man auf dem in § 15 eingeschlagenen Wege mit Rücksicht auf das Verschwinden von

$$g_{\mu\nu}\frac{\partial g^{\mu\nu}}{\partial x_\sigma}$$

die Gleichung

$$\frac{\partial t_\sigma^\alpha}{\partial x_\alpha} + \frac{1}{2}\frac{\partial g^{\mu\nu}}{\partial x_\sigma}T_{\mu\nu} = 0,$$

oder mit Rücksicht auf (56)

$$\frac{\partial T_\sigma^\alpha}{\partial x_\alpha} + \frac{1}{2}\frac{\partial g^{\mu\nu}}{\partial x_\sigma}T_{\mu\nu} = 0. \tag{57}$$

Ein Vergleich mit (41b) zeigt, daß diese Gleichung bei der getroffenen Wahl für das Koordinatensystem nichts anderes aussagt als das Verschwinden der Divergenz des Tensors der Energiekomponenten der Materie. Physikalisch zeigt das Auftreten des zweiten Gliedes der linken Seite, daß für die Materie allein Erhaltungssätze des Impulses und der Energie im eigent-

Einstein. 4

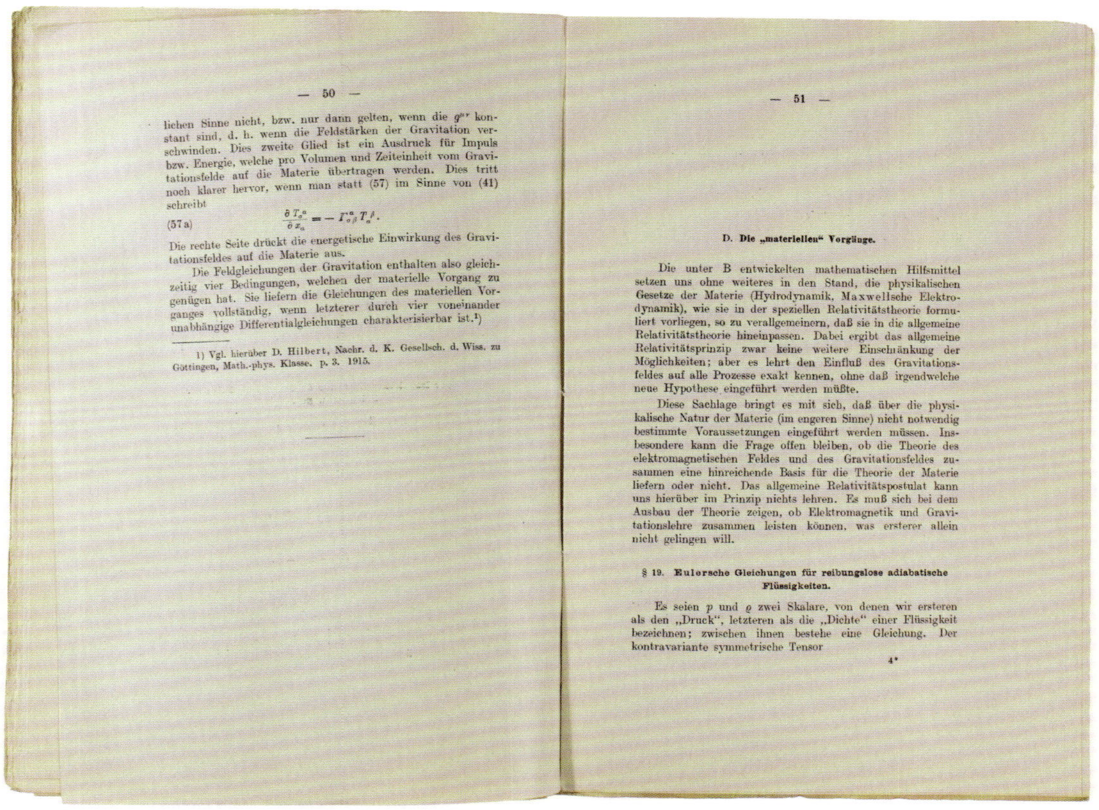

— 50 —

lichen Sinne nicht, bzw. nur dann gelten, wenn die $g^{\mu\nu}$ konstant sind, d. h. wenn die Feldstärken der Gravitation verschwinden. Dies zweite Glied ist ein Ausdruck für Impuls bzw. Energie, welche pro Volumen und Zeiteinheit vom Gravitationsfelde auf die Materie übertragen werden. Dies tritt noch klarer hervor, wenn man statt (57) im Sinne von (41) schreibt

$$\frac{\partial T_\sigma^\alpha}{\partial x_\alpha} = -\Gamma^{\alpha}_{\sigma\beta}T_\alpha^\beta. \tag{57a}$$

Die rechte Seite drückt die energetische Einwirkung des Gravitationsfeldes auf die Materie aus.

Die Feldgleichungen der Gravitation enthalten also gleichzeitig vier Bedingungen, welchen der materielle Vorgang zu genügen hat. Sie liefern die Gleichungen des materiellen Vorganges vollständig, wenn letzterer durch vier voneinander unabhängige Differentialgleichungen charakterisierbar ist.[1]

1) Vgl. hierüber D. Hilbert, Nachr. d. K. Gesellsch. d. Wiss. zu Göttingen, Math.-phys. Klasse. p. 3. 1915.

— 51 —

D. Die „materiellen" Vorgänge.

Die unter B entwickelten mathematischen Hilfsmittel setzen uns ohne weiteres in den Stand, die physikalischen Gesetze der Materie (Hydrodynamik, Maxwellsche Elektrodynamik), wie sie in der speziellen Relativitätstheorie formuliert vorliegen, so zu verallgemeinern, daß sie in die allgemeine Relativitätstheorie hineinpassen. Dabei ergibt das allgemeine Relativitätsprinzip zwar keine weitere Einschränkung der Möglichkeiten; aber es lehrt den Einfluß des Gravitationsfeldes auf alle Prozesse exakt kennen, ohne daß irgendwelche neue Hypothese eingeführt werden müßte.

Diese Sachlage bringt es mit sich, daß über die physikalische Natur der Materie (im engeren Sinne) nicht notwendig bestimmte Voraussetzungen eingeführt werden müssen. Insbesondere kann die Frage offen bleiben, ob die Theorie des elektromagnetischen Feldes und des Gravitationsfeldes zusammen eine hinreichende Basis für die Theorie der Materie liefern oder nicht. Das allgemeine Relativitätspostulat kann uns hierüber im Prinzip nichts lehren. Es muß sich bei dem Ausbau der Theorie zeigen, ob Elektromagnetik und Gravitationslehre zusammen leisten können, was ersterer allein nicht gelingen will.

§ 19. Eulersche Gleichungen für reibungslose adiabatische Flüssigkeiten.

Es seien $p$ und $\varrho$ zwei Skalare, von denen wir ersteren als den „Druck", letzteren als die „Dichte" einer Flüssigkeit bezeichnen; zwischen ihnen bestehe eine Gleichung. Der kontravariante symmetrische Tensor

4*

$$\left.\begin{aligned}&\delta\left\{\int H\,d\tau\right\}=0\\&H=g^{\mu\nu}\Gamma^{\alpha}_{\mu\beta}\Gamma^{\beta}_{\nu\alpha}\\&\sqrt{-g}=1\,.\end{aligned}\right\}\qquad(47\mathrm{a})$$

Dabei verschwinden die Variationen an den Grenzen des betrachteten begrenzten vierdimensionalen Integrationsraumes.

Es ist zunächst zu zeigen, daß die Form (47a) den Gleichungen (47) äquivalent ist. Zu diesem Zweck betrachten wir $H$ als Funktion der $g^{\mu\nu}$ und der

$$g^{\mu\nu}_{\sigma}\left(=\frac{\partial g^{\mu\nu}}{\partial x_{\sigma}}\right).$$

Dann ist zunächst

$$\begin{aligned}\delta H&=\Gamma^{\alpha}_{\mu\beta}\Gamma^{\beta}_{\nu\alpha}\delta g^{\mu\nu}+2g^{\mu\nu}\Gamma^{\alpha}_{\mu\beta}\,\delta\Gamma^{\beta}_{\nu\alpha}\\&=-\Gamma^{\alpha}_{\mu\beta}\Gamma^{\beta}_{\nu\alpha}\,\delta g^{\mu\nu}+2\,\Gamma^{\alpha}_{\mu\beta}\,\delta\left(g^{\mu\nu}\Gamma^{\beta}_{\nu\alpha}\right).\end{aligned}$$

Nun ist aber

$$\delta(g^{\mu\nu}\Gamma^{\beta}_{\nu\alpha})=-\tfrac{1}{2}\,\delta\left[g^{\mu\nu}g^{\beta\lambda}\left(\frac{\partial g_{\nu\lambda}}{\partial x_{\alpha}}+\frac{\partial g_{\alpha\lambda}}{\partial x_{\nu}}-\frac{\partial g_{\alpha\nu}}{\partial x_{\lambda}}\right)\right].$$

Die aus den beiden letzten Termen der runden Klammer hervorgehenden Terme sind von verschiedenem Vorzeichen und gehen auseinander (da die Benennung der Summationsindizes belanglos ist) durch Vertauschung der Indizes $\mu$ und $\beta$ hervor. Sie heben einander im Ausdruck für $\delta H$ weg, weil sie mit der bezüglich der Indizes $\mu$ und $\beta$ symmetrischen Größe $\Gamma^{\alpha}_{\mu\beta}$ multipliziert werden. Es bleibt also nur das erste Glied der runden Klammer zu berücksichtigen, so daß man mit Rücksicht auf (31) erhält

$$\delta H=-\Gamma^{\alpha}_{\mu\beta}\Gamma^{\beta}_{\nu\alpha}\,\delta g^{\mu\nu}-\Gamma^{\alpha}_{\mu\beta}\,\delta g^{\mu\beta}_{\alpha}\,.$$

Es ist also

$$\left.\begin{aligned}&\frac{\partial H}{\partial g^{\mu\nu}}=-\Gamma^{\alpha}_{\mu\beta}\Gamma^{\beta}_{\nu\alpha}\\&\frac{\partial H}{\partial g^{\mu\nu}_{\sigma}}=\Gamma^{\sigma}_{\mu\nu}\,.\end{aligned}\right\}\qquad(48)$$

Die Ausführung der Variation in (47a) ergibt zunächst das Gleichungssystem

$$\frac{\partial}{\partial x_\alpha}\left(\frac{\partial H}{\partial g^{\mu\nu}_\alpha}\right) - \frac{\partial H}{\partial g^{\mu\nu}} = 0, \tag{47b}$$

welches wegen (48) mit (47) übereinstimmt, was zu beweisen war. — Multipliziert man (47b) mit $g^{\mu\nu}_\sigma$, so erhält man, weil

$$\frac{\partial g^{\mu\nu}_\sigma}{\partial x_\alpha} = \frac{\partial g^{\mu\nu}_\alpha}{\partial x_\sigma}$$

und folglich

$$g^{\mu\nu}_\sigma \frac{\partial}{\partial x_\alpha}\left(\frac{\partial H}{\partial g^{\mu\nu}_\alpha}\right) = \frac{\partial}{\partial x_\alpha}\left(g^{\mu\nu}_\sigma \frac{\partial H}{\partial g^{\mu\nu}_\alpha}\right) - \frac{\partial H}{\partial g^{\mu\nu}_\alpha}\frac{\partial g^{\mu\nu}_\alpha}{\partial x_\sigma}$$

die Gleichung

$$\frac{\partial}{\partial x_\alpha}\left(g^{\mu\nu}_\sigma \frac{\partial H}{\partial g^{\mu\nu}_\alpha}\right) - \frac{\partial H}{\partial x_\sigma} = 0$$

oder [1])

$$\left\{\begin{aligned} &\frac{\partial t^\alpha_\sigma}{\partial x_\alpha} = 0 \\ &-2\varkappa t^\alpha_\sigma = g^{\mu\nu}_\sigma \frac{\partial H}{\partial g^{\mu\nu}_\alpha} - \delta^\alpha_\sigma H, \end{aligned}\right. \tag{49}$$

oder, wegen (48), der zweiten Gleichung (47) und (34)

$$\varkappa t^\alpha_\sigma = \tfrac{1}{2}\delta^\alpha_\sigma g^{\mu\nu} \Gamma^\lambda_{\mu\beta}\Gamma^\beta_{\nu\lambda} - g^{\mu\nu}\Gamma^\alpha_{\mu\beta}\Gamma^\beta_{\nu\sigma}. \tag{50}$$

Es ist zu beachten, daß $t_\sigma^\alpha$ kein Tensor ist; dagegen gilt (49) für alle Koordinatensysteme, für welche $\sqrt{-g} = 1$ ist. Diese Gleichung drückt den Erhaltungssatz des Impulses und der Energie für das Gravitationsfeld aus. In der Tat liefert die Integration dieser Gleichung über ein *dreidimensionales* Volumen $V$ die vier Gleichungen

$$\frac{d}{d x_4}\left\{\int t_\sigma^4 \, dV\right\} = \int (t_\sigma^1 \alpha_1 + t_\sigma^2 \alpha_2 + t_\sigma^3 \alpha_3)\, dS, \tag{49a}$$

wobei $\alpha_1$, $\alpha_2$, $\alpha_3$ der Richtungskosinus der nach innen gerichteten Normale eines Flächenelementes der Begrenzung

1) Der Grund der Einführung des Faktors $-2\varkappa$ wird später deutlich werden.

### *THE THEORETICAL BASIS FOR THE LASER*

Einstein published two papers, just weeks apart in 1916, in which he proposed that atomic radiation emission—which had been discovered to occur spontaneously—could also be produced artificially.

In the first paper, *Emission and Absorption of Radiation in Quantum Theory*, Einstein wrestled with the question of whether atomic radiation emission could be directed (controlled and focused). He explored this concept further in his second paper, *On the Quantum Theory of Radiation*. In the latter paper, he convincingly demonstrated that atomic radiation could indeed be directed.

As usual, Einstein proposed a thought experiment. He envisioned an atom-filled chamber bathed in light. By combining Niels Bohr's model of the atom with Planck's theory of the quanta, Einstein proposed a theory of stimulated emission. His idea was that if a photon were to hit an electron in a high-energy state, the electron would drop to a lower energy state, releasing energy in the form of a new, identical proton.

Spontaneous emission, or radioactive decay, occurs randomly and without external force. On the other hand, stimulated emission is predictable, and the radiation that it produces has characteristics (frequency, polarization, and direction of travel) identical to the incident energy source. By controlling and repeating the process, stimulated emission can produce large quantities of identical photons. When released as a coherent beam, all photons have parallel rays and identical wavelengths. The result is highly amplified light with much greater intensity than ordinary light.

ALBERT EINSTEIN, Strahlungs-Emission und -Absorption nach der Quantentheorie (Emission and Absorption of Radiation in Quantum Theory) *and* Zur Quantentheorie der Strahlung (On the Quantum Theory of Radiation), *1916. Uncommon offprints of Einstein's work on quantum theory first published in Mitteilungen der Physikalischen Gesellschaft Zurich.*

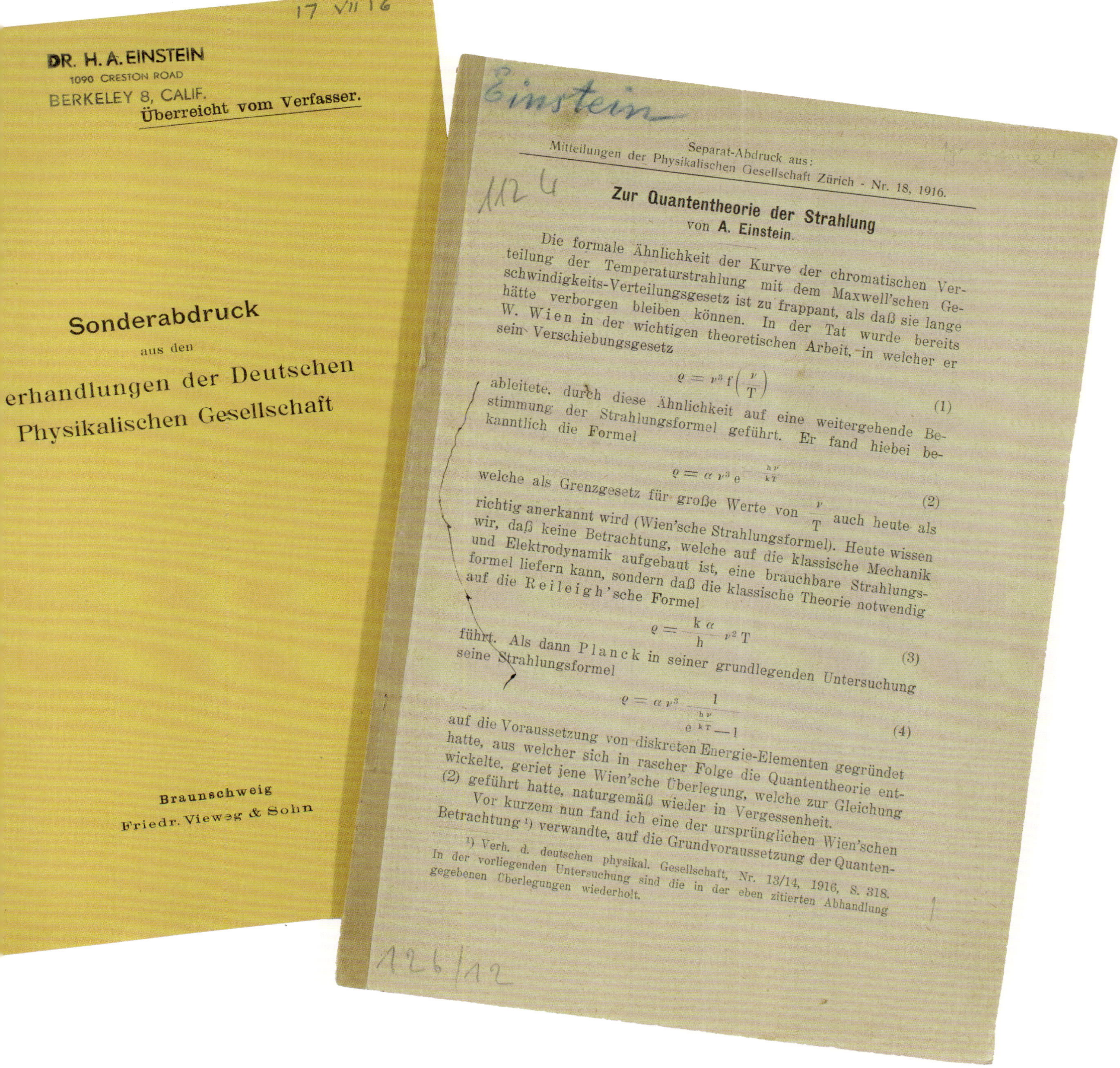

DR. H. A. EINSTEIN
1090 CRESTON ROAD
BERKELEY 8, CALIF.
Überreicht vom Verfasser.

**Sonderabdruck**
aus den
[V]erhandlungen der Deutschen
Physikalischen Gesellschaft

Braunschweig
Friedr. Vieweg & Sohn

Separat-Abdruck aus:
Mitteilungen der Physikalischen Gesellschaft Zürich - Nr. 18, 1916.

**Zur Quantentheorie der Strahlung**
von **A. Einstein.**

Die formale Ähnlichkeit der Kurve der chromatischen Verteilung der Temperaturstrahlung mit dem Maxwell'schen Geschwindigkeits-Verteilungsgesetz ist zu frappant, als daß sie lange hätte verborgen bleiben können. In der Tat wurde bereits W. Wien in der wichtigen theoretischen Arbeit, in welcher er sein Verschiebungsgesetz

$$\varrho = \nu^3 f\left(\frac{\nu}{T}\right) \qquad (1)$$

ableitete, durch diese Ähnlichkeit auf eine weitergehende Bestimmung der Strahlungsformel geführt. Er fand hiebei bekanntlich die Formel

$$\varrho = \alpha \nu^3 e^{-\frac{h\nu}{kT}} \qquad (2)$$

welche als Grenzgesetz für große Werte von $\frac{\nu}{T}$ auch heute als richtig anerkannt wird (Wien'sche Strahlungsformel). Heute wissen wir, daß keine Betrachtung, welche auf die klassische Mechanik und Elektrodynamik aufgebaut ist, eine brauchbare Strahlungsformel liefern kann, sondern daß die klassische Theorie notwendig auf die Reileigh'sche Formel

$$\varrho = \frac{k\alpha}{h}\nu^2 T \qquad (3)$$

führt. Als dann Planck in seiner grundlegenden Untersuchung seine Strahlungsformel

$$\varrho = \alpha \nu^3 \frac{1}{e^{\frac{h\nu}{kT}} - 1} \qquad (4)$$

auf die Voraussetzung von diskreten Energie-Elementen gegründet hatte, aus welcher sich in rascher Folge die Quantentheorie entwickelte, geriet jene Wien'sche Überlegung, welche zur Gleichung (2) geführt hatte, naturgemäß wieder in Vergessenheit.

Vor kurzem nun fand ich eine der ursprünglichen Wien'schen Betrachtung [1]) verwandte, auf die Grundvoraussetzung der Quanten-

[1]) Verh. d. deutschen physikal. Gesellschaft, Nr. 13/14, 1916, S. 318. In der vorliegenden Untersuchung sind die in der eben zitierten Abhandlung gegebenen Überlegungen wiederholt.

$$R_{\mu\nu} - \frac{1}{2} g_{\mu\nu} R = -T_{\mu\nu}$$

## *ON GRAVITATIONAL WAVES*

According to the general theory of relativity, gravity is the result of the universe's curvature, which is affected by the mass of the bodies within it. The universe's curvature determines the path that objects travel. Furthermore, it is flexible and bends as those objects move.

Einstein's theory predicted that accelerating objects would produce gravitational waves radiating in all directions at the speed of light.

Einstein initially published his equations for gravitational waves in 1916. Realizing that he had made an error, he published a second paper in 1918 (shown here) that corrected the error and elaborated on the ideas of his previous paper. Still, Einstein had only approximated the energy of gravitational waves. He returned to this subject decades later and in 1937 published an exact formula for gravitational radiation.

Einstein believed that gravitational waves must exist, but because their intensity would be so low, he thought they would never be detected.

Proof of their existence came in 1974, two decades after Einstein's death. Astronomers at the Arecibo Radio Observatory in Puerto Rico observed a binary pulsar for over eight years. They found that the two stars were moving closer to each other at precisely the rate they would if they were radiating gravitational waves.

Four decades after that, with the development of the Laser Interferometer Gravitational-Wave Observatory (LIGO), direct detection of gravitational waves from the merger of two black holes was reported on September 14, 2015. Since then, gravitational waves have been recorded from the merger of numerous pairs of black holes, neutron stars, and combinations of black holes and neutron stars.

Before LIGO, astronomers could only observe the universe using light signals (the entire electromagnetic spectrum, not only visible light). The ability to measure gravitational radiation is a profound development in our attempts to perceive the cosmos. It's as if the observable universe can now be heard as well as seen.

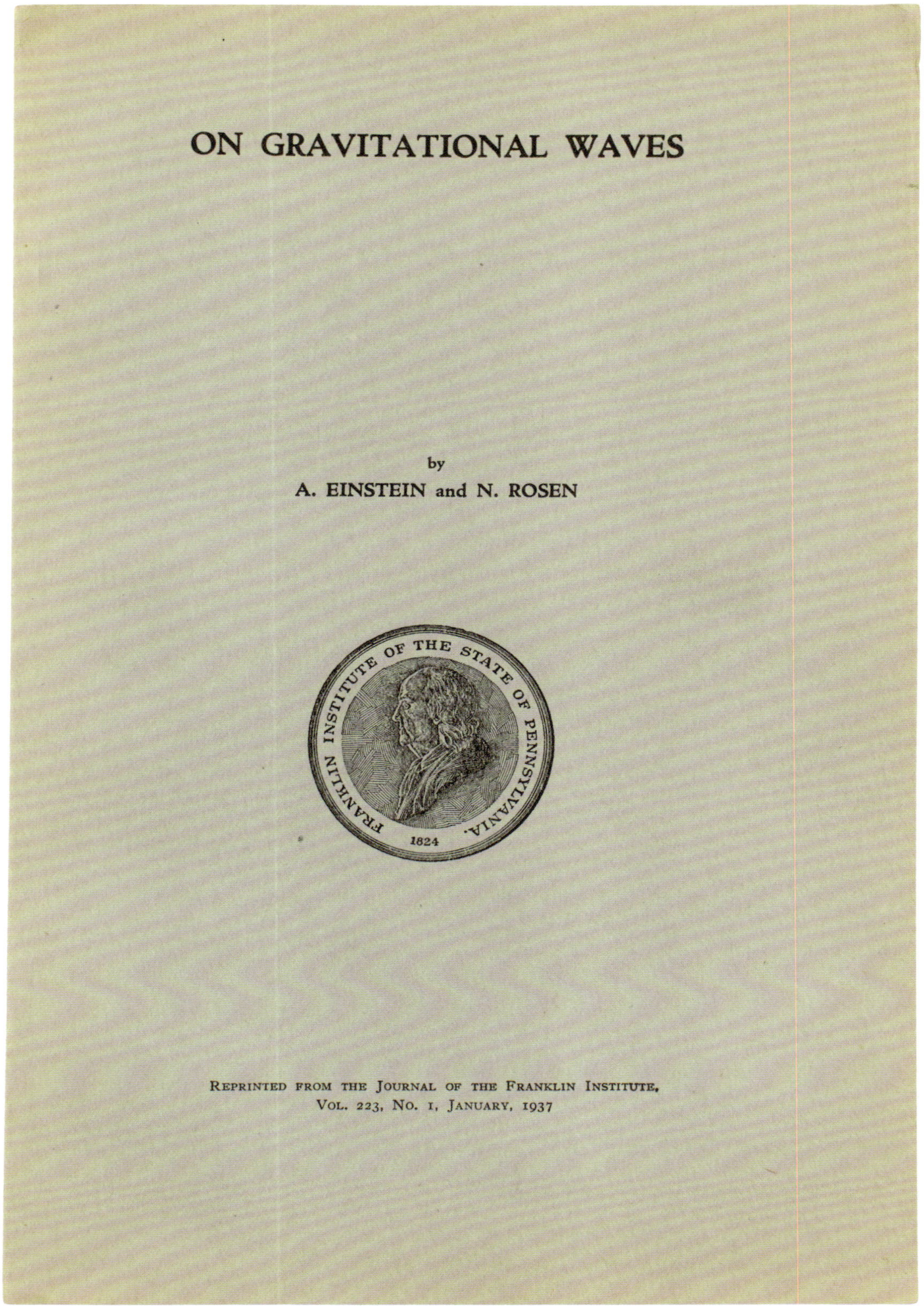

ON GRAVITATIONAL WAVES

by
A. EINSTEIN and N. ROSEN



REPRINTED FROM THE JOURNAL OF THE FRANKLIN INSTITUTE,
VOL. 223, NO. 1, JANUARY, 1937

*Above:* ALBERT EINSTEIN AND NATHAN ROSEN, On Gravitational Waves, *1937. First offprint from* The Journal of the Franklin Institute. *Opposite:* ALBERT EINSTEIN, *"Über Gravitationswellen" ("On Gravitational Waves") in* Sitzungsberichte der Königlich Preussischen Akademie der Wissenschaften zu Berlin (Reports from the Prussian Academy of Sciences in Berlin), *1918.*

SITZUNGSBERICHTE 1918.
VIII.

DER

KÖNIGLICH PREUSSISCHEN

AKADEMIE DER WISSENSCHAFTEN.

Gesamtsitzung vom 14. Februar.
Mitteilung vom 31. Januar.

# Über Gravitationswellen.

Von A. Einstein.

Sonderabdruck.

Verlag der Königlichen Akademie der Wissenschaften.
In Kommission bei Georg Reimer.

(Preis ℳ 0.50)

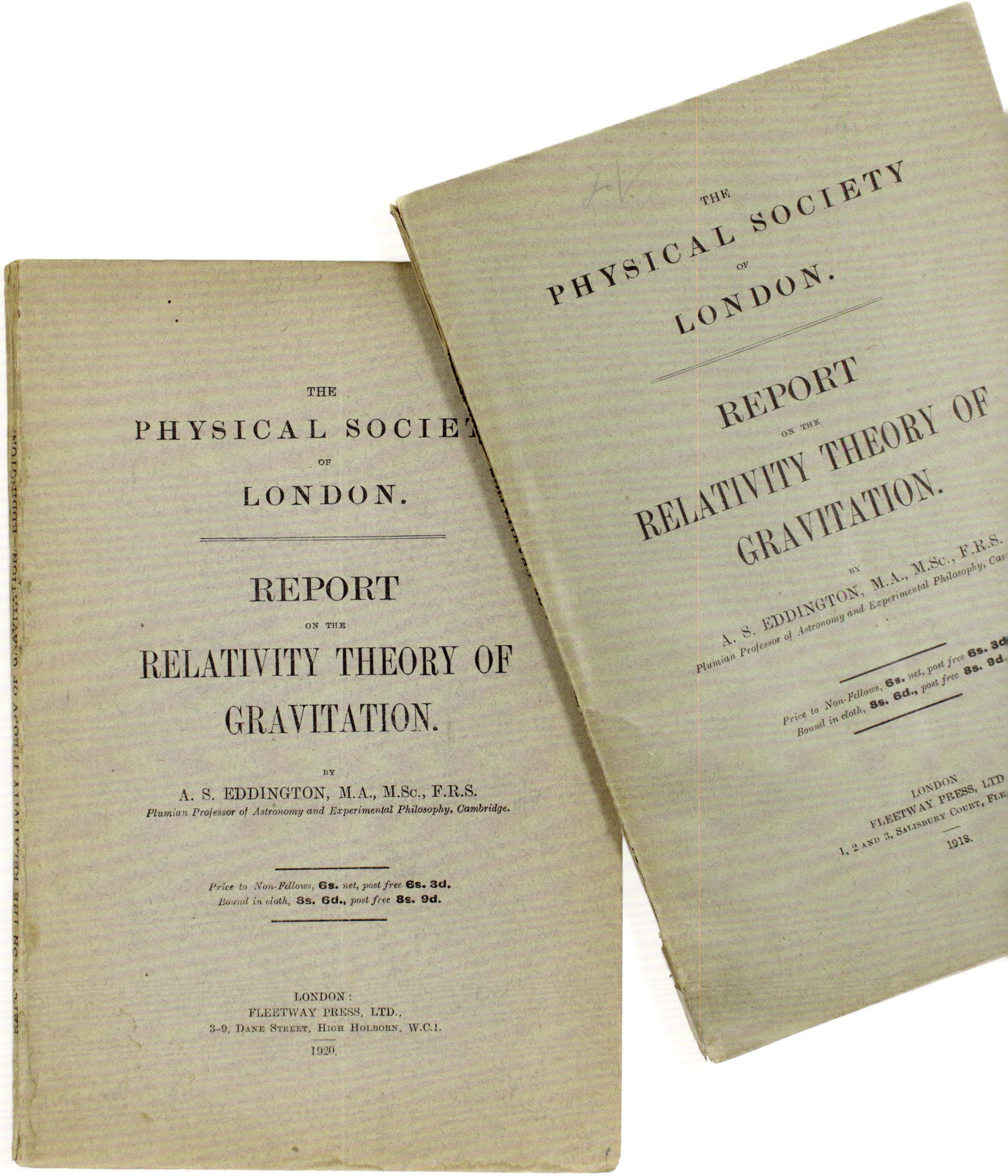
THE
PHYSICAL SOCIETY
OF
LONDON.
REPORT
ON THE
RELATIVITY THEORY OF
GRAVITATION.
BY
A. S. EDDINGTON, M.A., M.Sc., F.R.S.
Plumian Professor of Astronomy and Experimental Philosophy, Cambridge.
Price to Non-Fellows, 6s. net, post free 6s. 3d.
Bound in cloth, 8s. 6d., post free 8s. 9d.
LONDON:
FLEETWAY PRESS, LTD.,
3–9, DANE STREET, HIGH HOLBORN, W.C.1.
1920.
THE
PHYSICAL SOCIETY
OF
LONDON.
REPORT
ON THE
RELATIVITY THEORY OF
GRAVITATION.
BY
A. S. EDDINGTON, M.A., M.Sc., F.R.S.
LONDON
FLEETWAY PRESS, LTD
1918.

### *FIRST EXPERIMENTAL TEST OF GENERAL RELATIVITY*

British astronomer Sir Arthur Eddington provided the first experimental proof, in 1919, of the general theory of relativity. Until then, Einstein's revolutionary theory was little known or understood, even among physicists.

The first edition of Eddington's report on general relativity, from 1918, introduced Einstein's theory to the English-speaking world. The second edition, from 1920, has a preface discussing the results of Eddington's famous eclipse expedition to the West African island of Príncipe in May 1919.

Einstein's theory proposed that gravity resulted from the curvature of spacetime by mass. It predicted precisely how light rays would bend as they traveled along the curvature.

Eddington tested Einstein's theory by measuring the deflection of starlight by the Sun's gravity during a total solar eclipse. He compared the apparent positions of stars near the Sun (visible in the daytime because of the eclipse) to their positions as determined at night (when their light does not pass by the Sun).

Einstein's theory predicted that light would be deflected twice the amount predicted by Newton's law of universal gravitation. Eddington's measurements proved Einstein's prediction, not Newton's, to be correct.

Eddington presented his expedition results to the Royal Society and the Royal Astronomical Society in London on November 6, 1919. In the following days, newspapers worldwide ran headline stories accompanied by Einstein's photo. Because of Eddington, Einstein became an international celebrity practically overnight.

A.S. EDDINGTON, Report on the Relativity Theory of Gravitation, *1918 and 1920.*

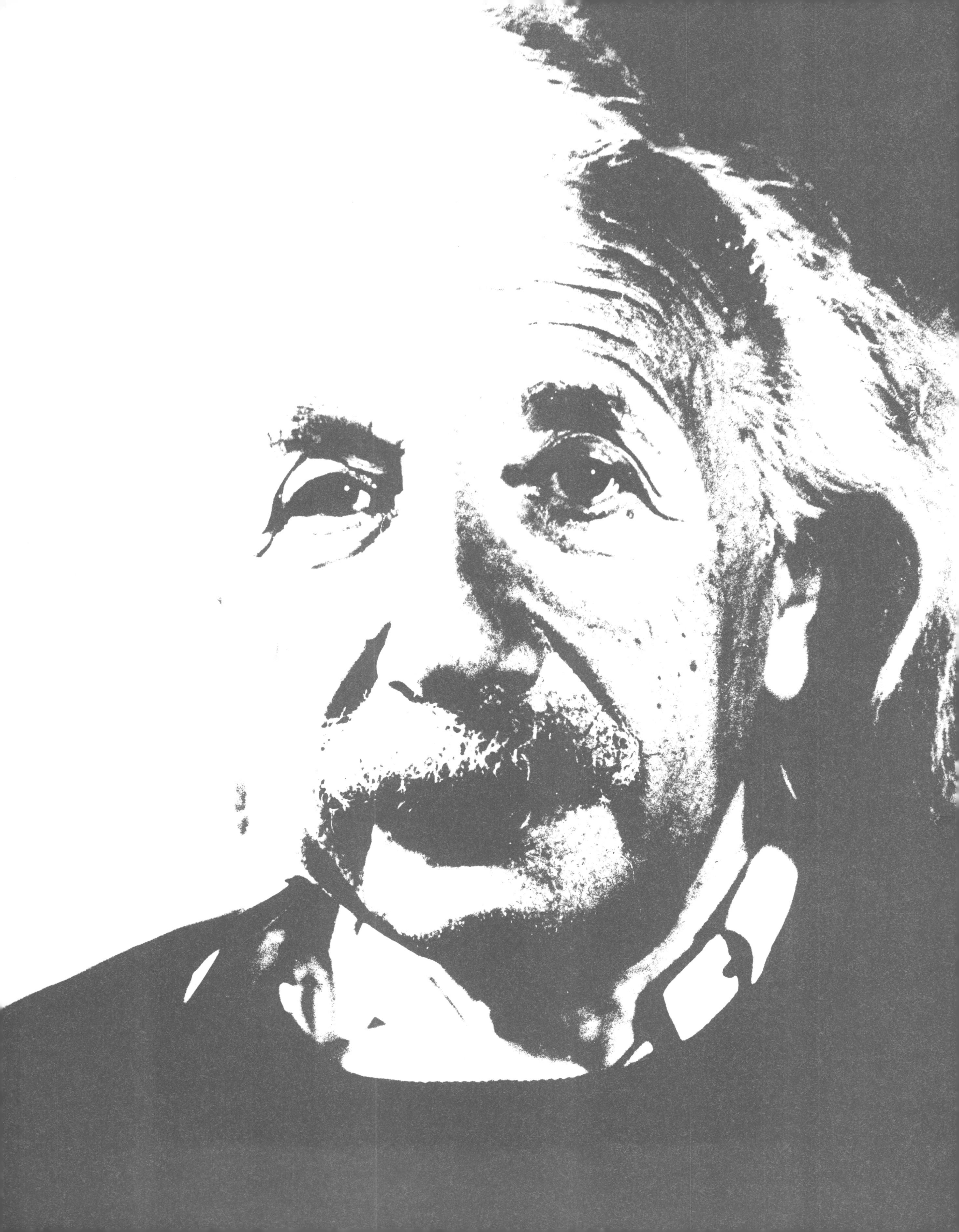

*The non-mathematician is seized by a mysterious shuddering when he hears of 'four dimensional' things, by a feeling not unlike that awakened by thoughts of the occult. And yet there is no more common-place statement than that the world in which we live is a four-dimensional space-time continuum."*

—ALBERT EINSTEIN, *1920*

## *RELATIVITY EXPLAINED TO THE PUBLIC*

Einstein's most widely read publication, *Relativity: The Special and the General Theory*, presents a simplified account of relativity. Although Einstein "found it difficult to write at this level, he felt he had no choice but to do so if his theories were to be understood. . . . The book was a huge success. . . . Translations into other languages followed, making relativity theory known throughout the world."

By 1920, when Einstein published this book, the difficulty in understanding relativity theory had become the subject of humor. In an apocryphal story, when Sir Arthur Eddington was asked how it felt to be one of only three people in the world to comprehend relativity, he paused in responding, leading his interviewer to ask, "What is wrong, Mr. Eddington?" Eddington replied: "I'm sorry, I was just wondering who the third person is."

*Relativity: The Special and the General Theory* was Einstein's extremely successful effort to make relativity theory understandable to the public.

*Pages 51–53:* ALBERT EINSTEIN, Relativity: The Special and the General Theory, *1920. First edition.*

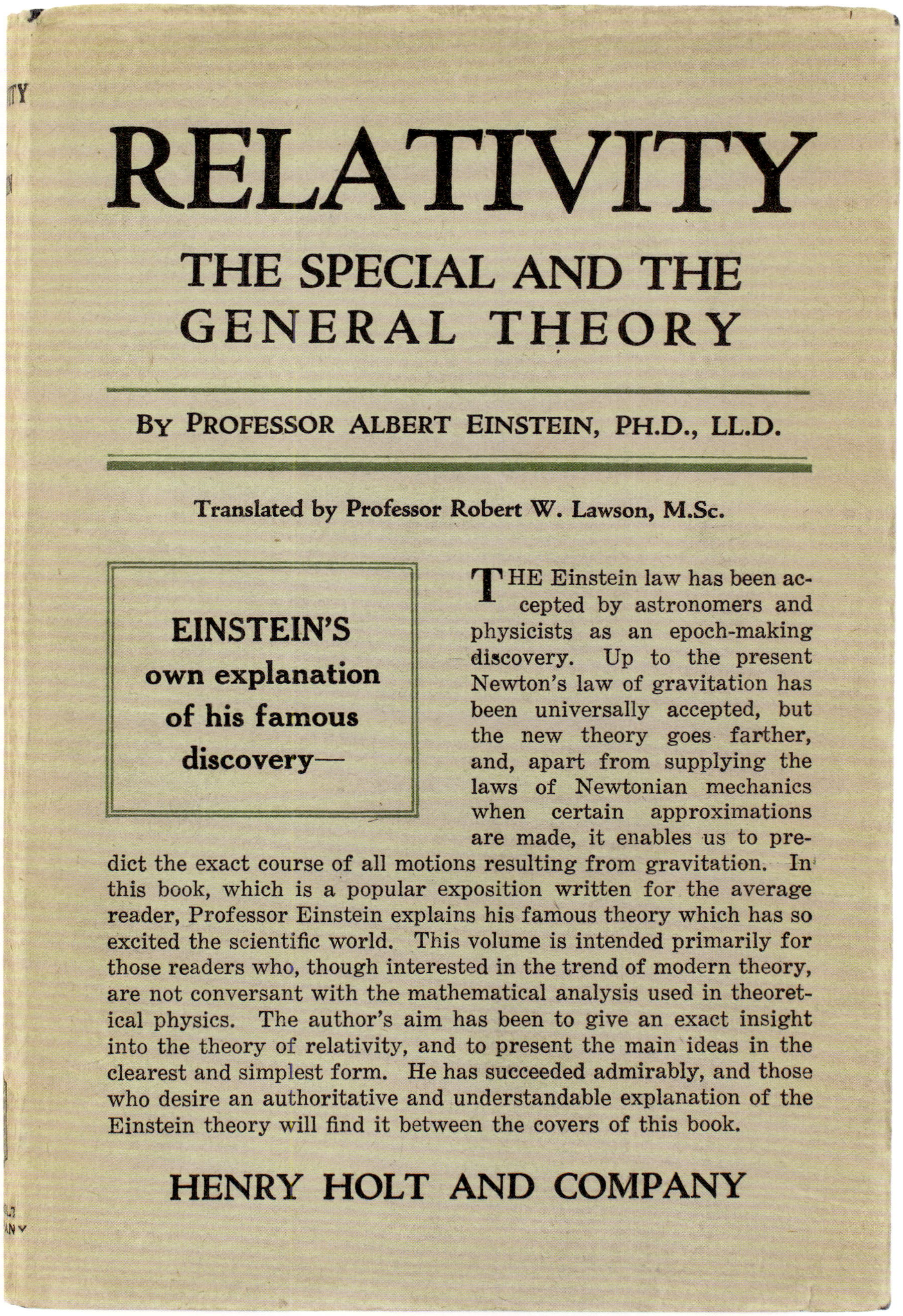

# RELATIVITY

## THE SPECIAL AND THE GENERAL THEORY

**By PROFESSOR ALBERT EINSTEIN, PH.D., LL.D.**

**Translated by Professor Robert W. Lawson, M.Sc.**

**EINSTEIN'S own explanation of his famous discovery—**

THE Einstein law has been accepted by astronomers and physicists as an epoch-making discovery. Up to the present Newton's law of gravitation has been universally accepted, but the new theory goes farther, and, apart from supplying the laws of Newtonian mechanics when certain approximations are made, it enables us to predict the exact course of all motions resulting from gravitation. In this book, which is a popular exposition written for the average reader, Professor Einstein explains his famous theory which has so excited the scientific world. This volume is intended primarily for those readers who, though interested in the trend of modern theory, are not conversant with the mathematical analysis used in theoretical physics. The author's aim has been to give an exact insight into the theory of relativity, and to present the main ideas in the clearest and simplest form. He has succeeded admirably, and those who desire an authoritative and understandable explanation of the Einstein theory will find it between the covers of this book.

**HENRY HOLT AND COMPANY**

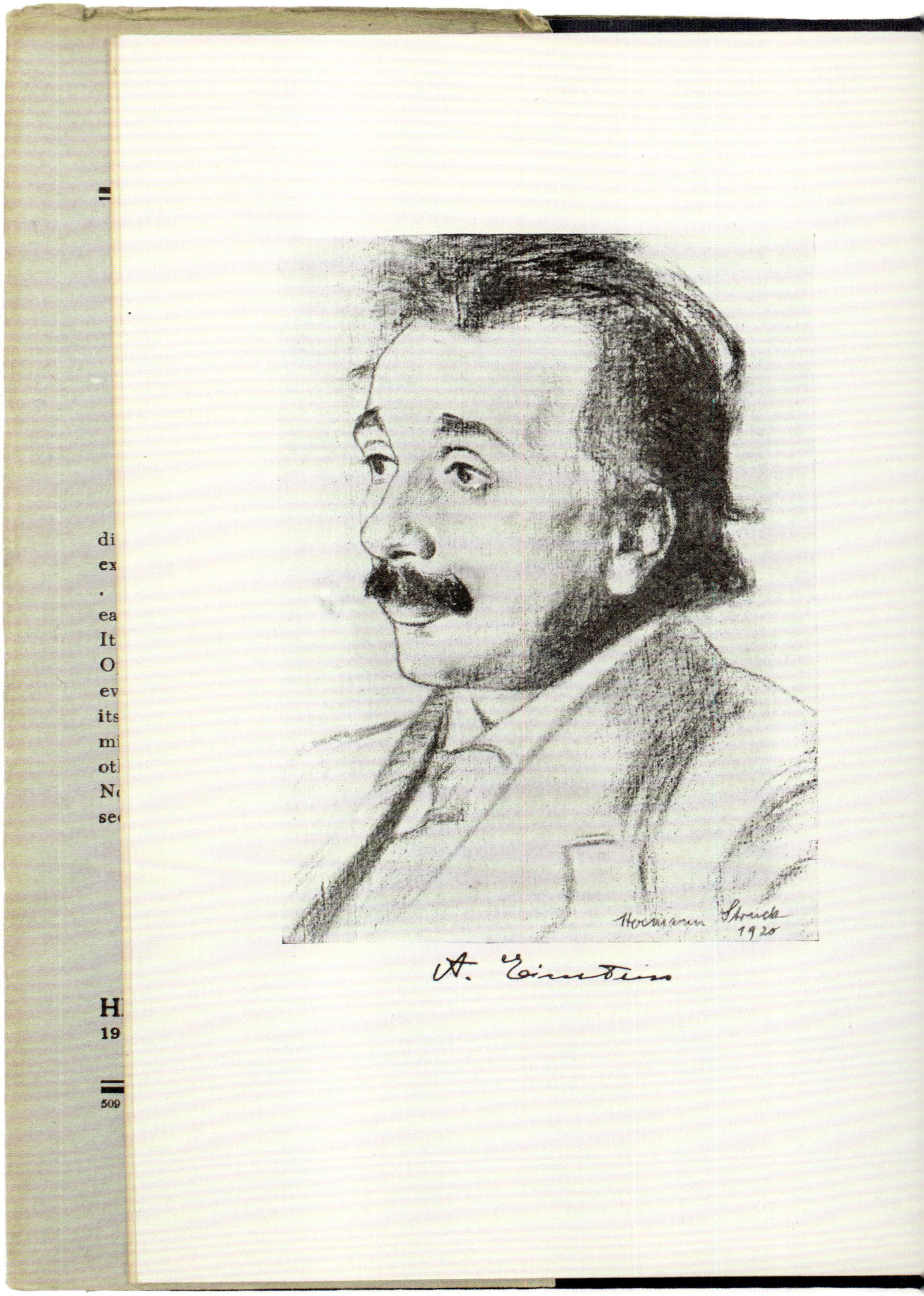
Hermann Struck
1920

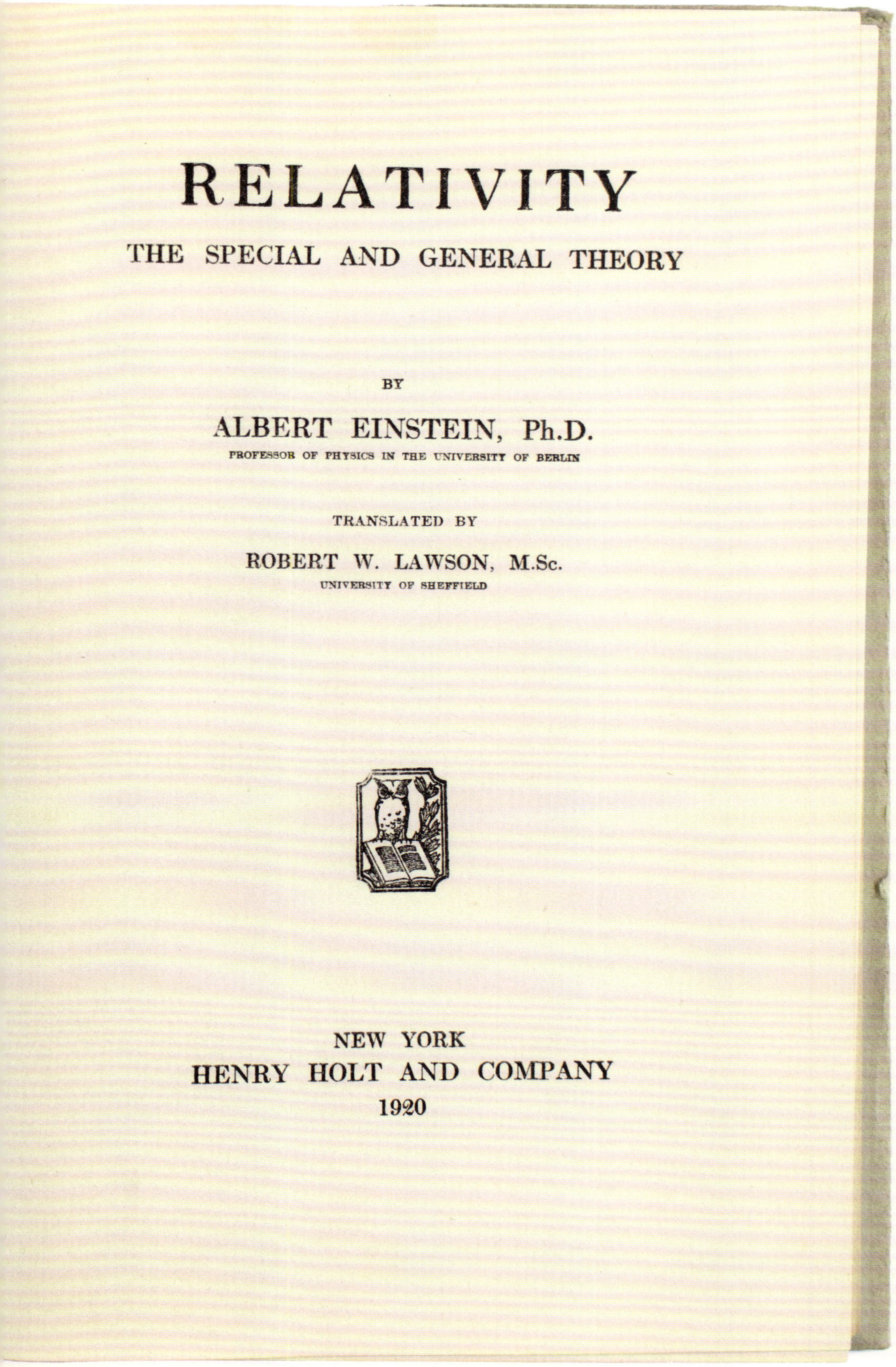

# RELATIVITY

## THE SPECIAL AND GENERAL THEORY

BY

ALBERT EINSTEIN, Ph.D.

PROFESSOR OF PHYSICS IN THE UNIVERSITY OF BERLIN

TRANSLATED BY

ROBERT W. LAWSON, M.Sc.

UNIVERSITY OF SHEFFIELD

NEW YORK

HENRY HOLT AND COMPANY

1920

$$x' = \frac{(c-v)t}{\sqrt{1-\frac{v^2}{c^2}}}$$

$$t' = \frac{\left(1-\frac{v}{c}\right)t}{\sqrt{1-\frac{v^2}{c^2}}}$$

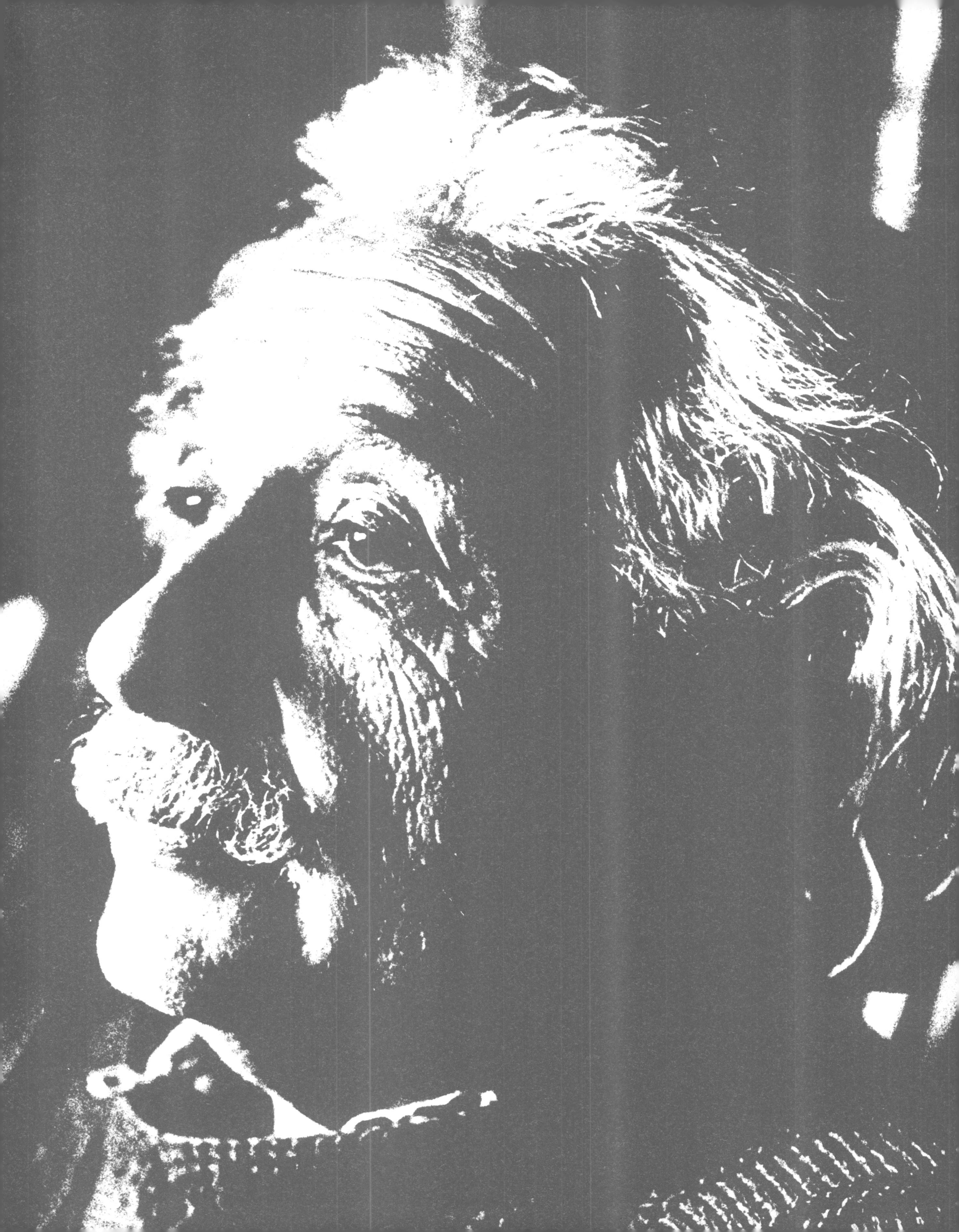

### *EINSTEIN IN ENGLAND*

Walter Benington's "famous photograph of Albert Einstein was taken in June 1921, when Einstein came to London, at the invitation of Lord Haldane, to deliver an important lecture on his theories of Relativity. Einstein's visit captured the public's interest. *The Sphere* (18 June 1921) reproduced Benington's informal double portrait of Einstein and Haldane as its front cover with an inset of Einstein taken at the same sitting."

This image is one of at least two that Benington took of Einstein by himself. Einstein likely received only a few prints at the time of the sitting for presentation purposes. He presented this photo to Ruth Blumgart.

Blumgart was an extremely close associate of Sigmund Freud. "Ruth's access to Freud seemed unique; she came to meals at his apartment, visited him in summers, and was on excellent terms with his children. . . . Ruth was considered a member of Freud's extended family." Einstein's German-language inscription on the photo reads: "To the lovely Mrs. Ruth Blumgart / Daughter of Judge Mack to whom I send fondest / Regards / Albert Einstein, July 1921."

Ruth's father, Julian W. Mack, was a distinguished federal judge and a leader of the American Jewish community. After immigrating to the United States, Einstein regularly visited Mack at his home. Einstein and Mack helped victims of Nazism, including many prominent scientists, escape Germany and receive academic appointments in the US.

The photograph is of exceptionally high quality, its soft tint enhancing Einstein's delicate expression. It is the only example of the photograph we could locate.

PHOTOGRAPH BY WALTER BENINGTON, *England, 1921. Signed, dated, and inscribed by Einstein: "To the lovely Mrs. Ruth Blumgart."*

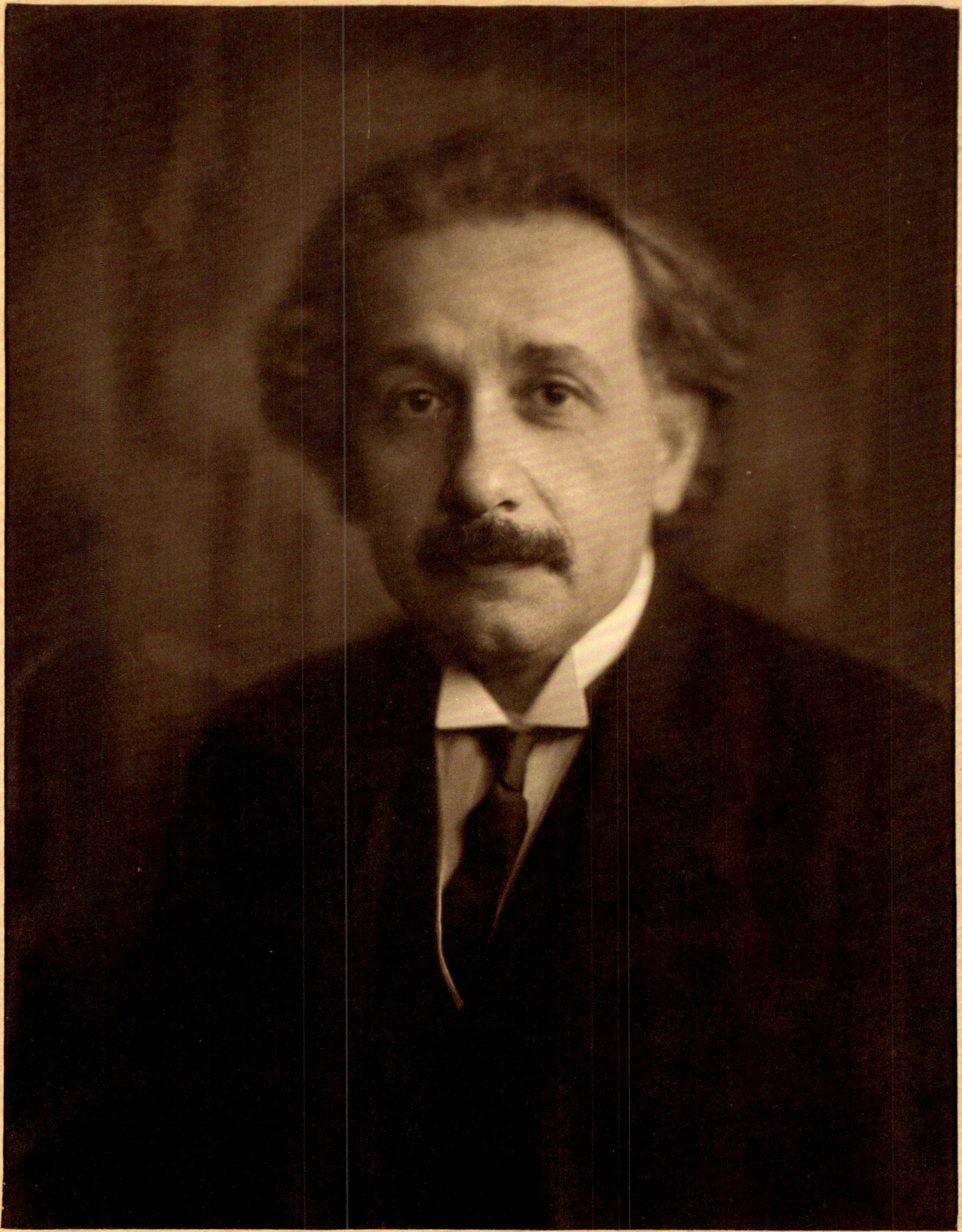

Frau Ruth Blumgart, der liebenswürdigen Tochter Judge Macks zum freundlichen Andenken

Albert Einstein. Juli 1921.

## *FRANCE IN THE SPRING*

Einstein traveled to Paris in the spring of 1922 to give a series of lectures about relativity at the Collège de France. During his trip, Einstein wrote to Sir Thomas Barclay and mentioned wanting to see him.

Sir Thomas Barclay was a celebrated liberal statesman, author, and expert on international relations and law. As an MP and deputy chairman of the International Law Association, he was a strong advocate for international peace.

Barclay invited Einstein to "share the midday meal" on April 1. Einstein later sent the photo shown here with an inscription that reads, translated, "Mr. Barclay in memory of 1 April 22 / Albert Einstein."

Seven months after this visit, in November 1922, Einstein was awarded the 1921 Nobel Prize in Physics.

This elegant portrait was taken by Henri Manuel. He and his brother, Gaston, operated the largest photographic studio in Paris at the time.

PHOTOGRAPH BY HENRI MANUEL, *Paris, France, 1922. Signed, dated, and inscribed by Einstein to Sir Thomas Barclay: "Mr. Barclay in memory of 1 April 22."*

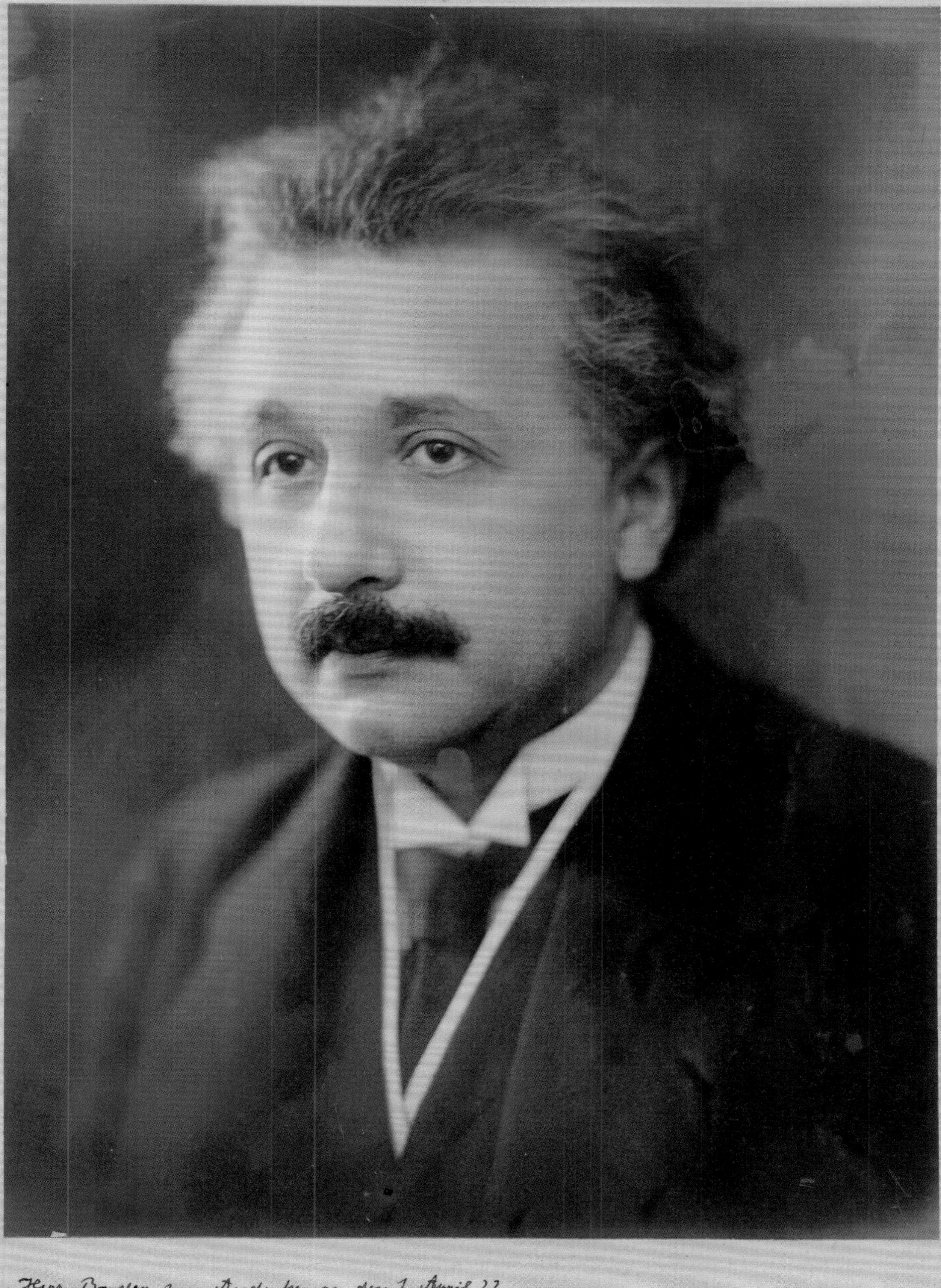

Herrn Barsley zum Andenken an den 1. April 22
Albert Einstein.

## *AN INNER VOICE*

This carte-de-visite-style photograph was designed for presentation. It shows forty-seven-year-old Einstein—the most highly regarded physicist in the world—at the height of his powers. He seems approachable, perhaps with a bit of mischief in his eyes.

The year 1926 was a pivotal one for Einstein. He started to become isolated from the scientific community over his philosophical refusal to accept the implications of the new quantum mechanics as developed in 1925–26 by Werner Heisenberg, Erwin Schrödinger, Pascual Jordan, Max Born, and others. In a letter to Max Born of December 4, 1926, Einstein expressed his reservations about the new theory:

"Quantum mechanics is very impressive. But an inner voice tells me that it is not yet the real thing. The theory produces a good deal but hardly brings us closer to the secret of the Old One."

UNKNOWN PHOTOGRAPHER,
*1926. Signed and dated by Einstein.*

*Quantum mechanics is very impressive. But an inner voice tells me that it is not yet the real thing. The theory produces a good deal but hardly brings us closer to the secret of the Old One."*

—ALBERT EINSTEIN, *1926*

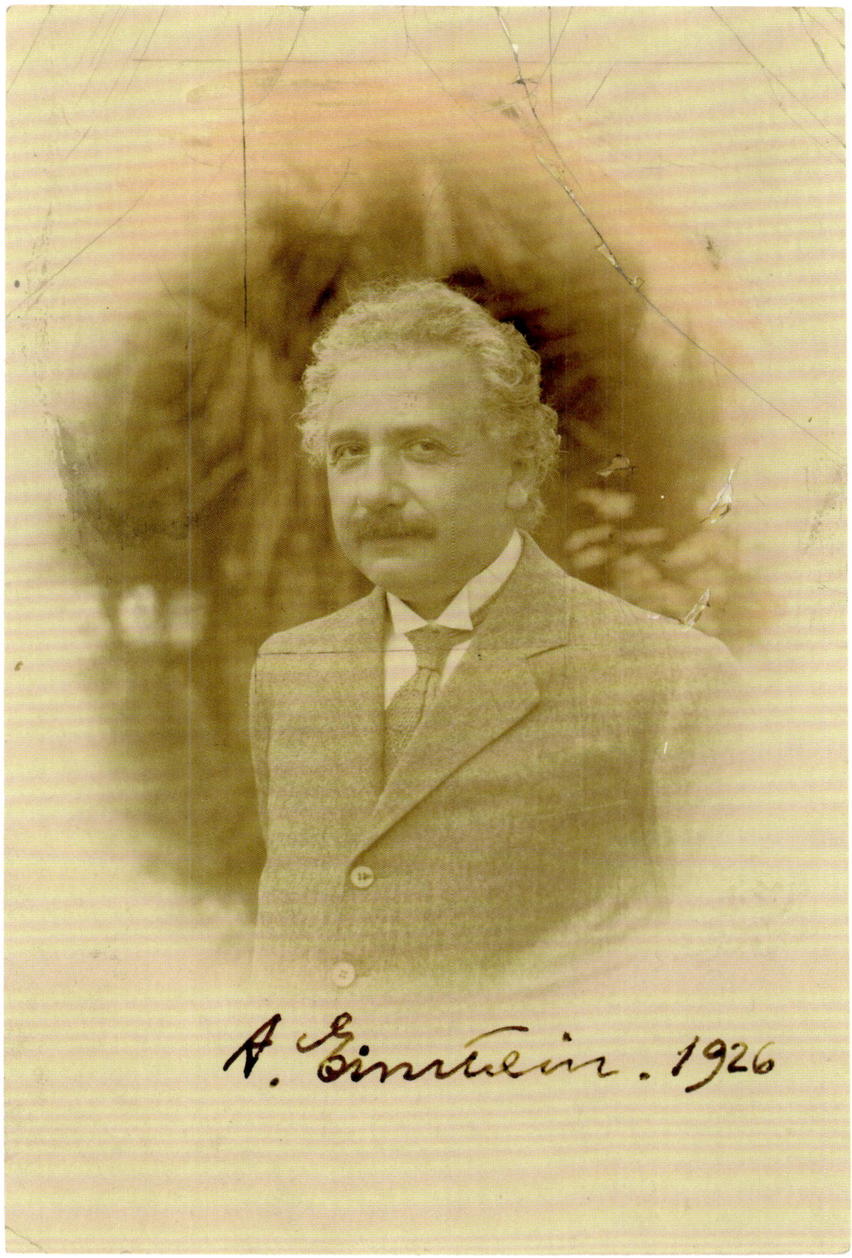
A. Einstein. 1926

## *GOD DOES NOT PLAY DICE*

This photograph captures the handsome, smartly attired Einstein in a thoughtful mood, customary pipe in hand. His bold signature on the mount reads: "Albert Einstein / Berlin September 1927." The photographer remains unidentified.

Just weeks after this image was taken, Einstein, age forty-eight, attended the most celebrated gathering of physicists of the twentieth century—the Fifth Solvay Conference—held in Belgium from October 24 to 29, 1927.

The conference was dominated by informal, spontaneous debates between Einstein and Niels Bohr, which continued as a lifelong dialogue between the two great physicists. Their debates were about the interpretation of quantum theory, a field that Einstein himself helped establish with his discovery of the dual particle/wave nature of light in 1905.

It was at the Fifth Solvay Conference where Einstein—disagreeing with Niels Bohr's probabilistic view of the universe—famously said, "God does not play dice," prompting Bohr's witty response, "Einstein, stop telling God what to do!"

UNKNOWN PHOTOGRAPHER,
*Berlin, Germany, 1927. Signed and dated by Einstein.*

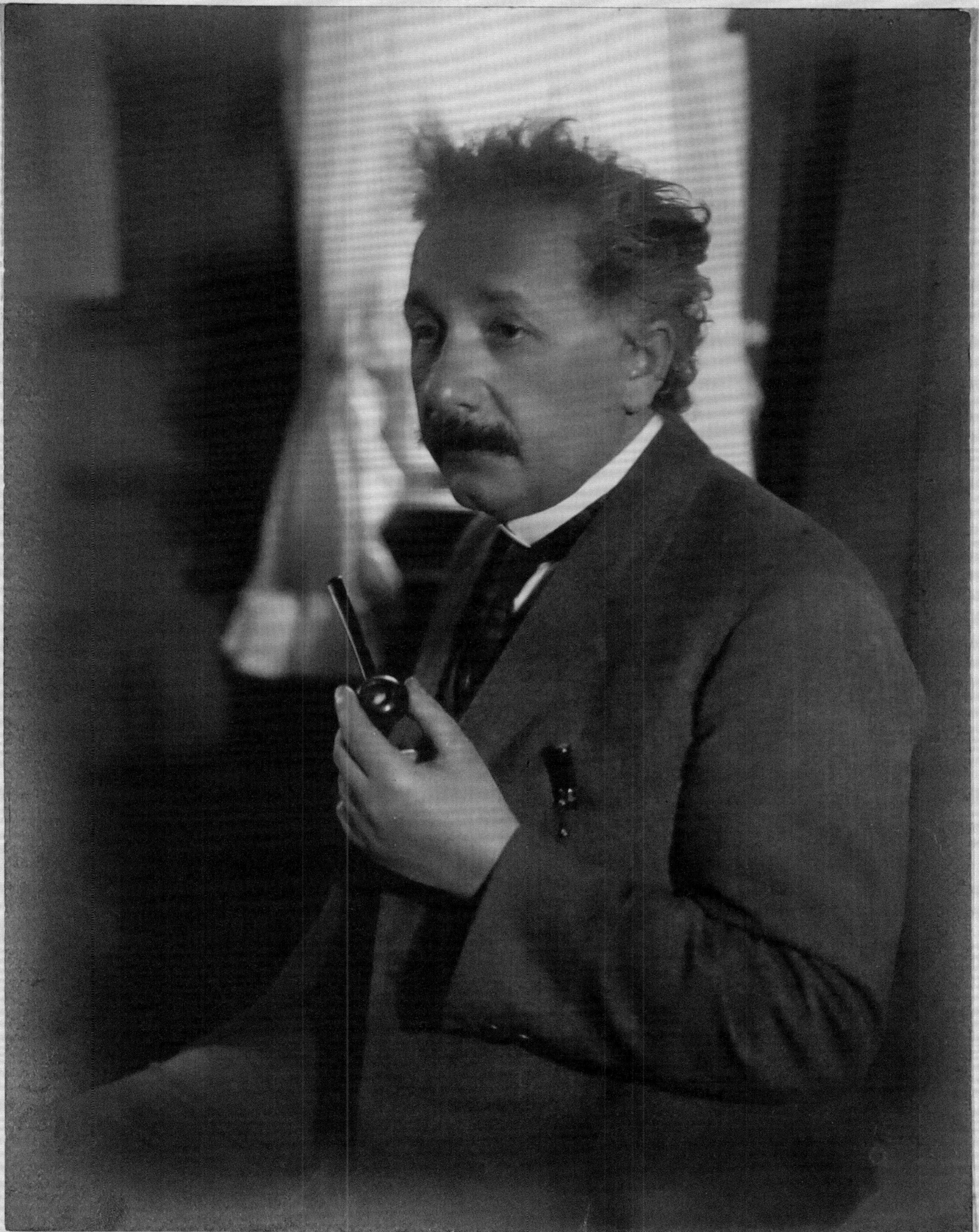

Albert Einstein

Berlin. September 1927.

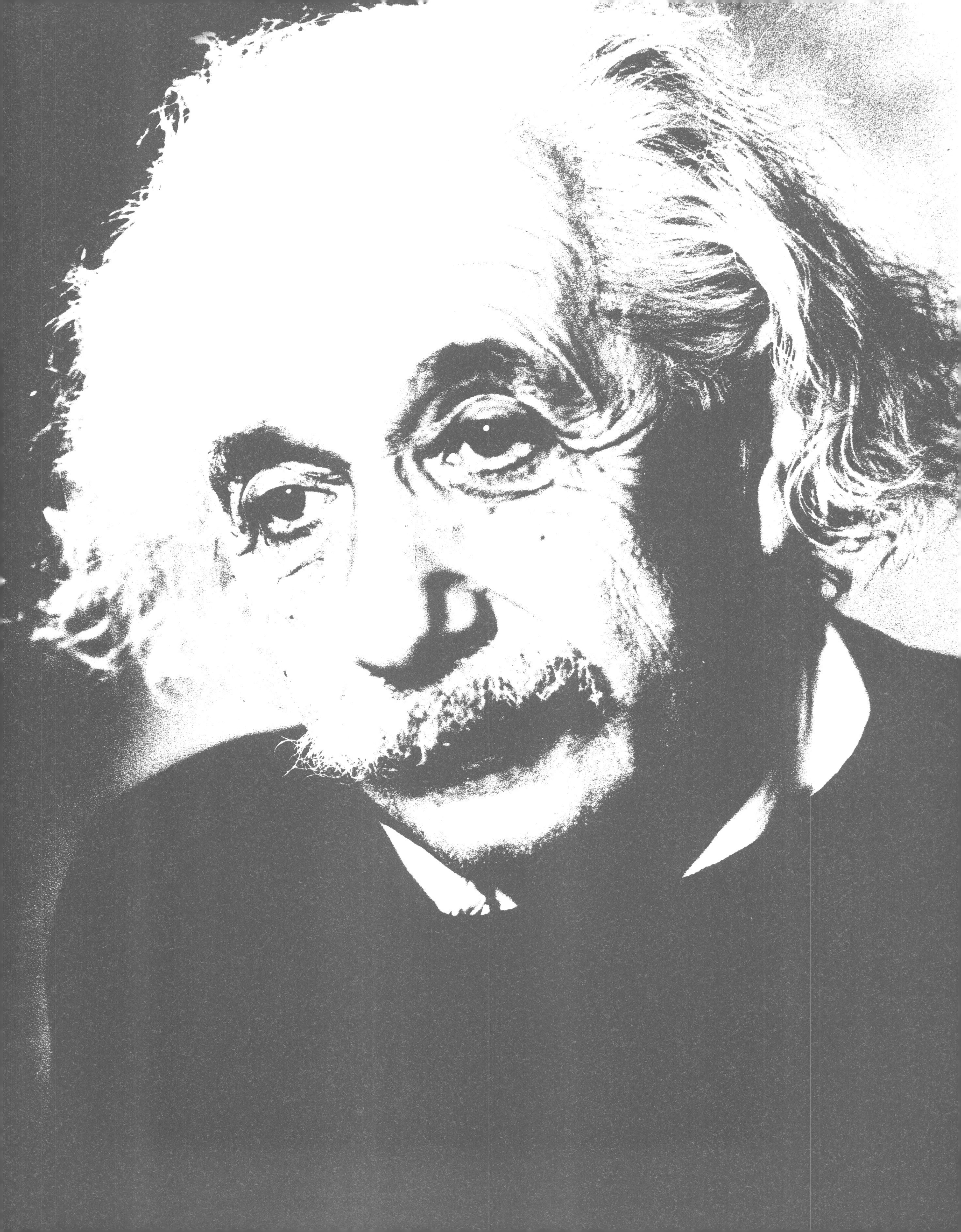

*Although personal illusions*
*may bring warmth*
*and joy into one's life*
*when young, they don't last*
*a lifetime. Life would be*
*bleak if the work and the passion*
*for discovery did not exist."*

—ALBERT EINSTEIN, *1937*

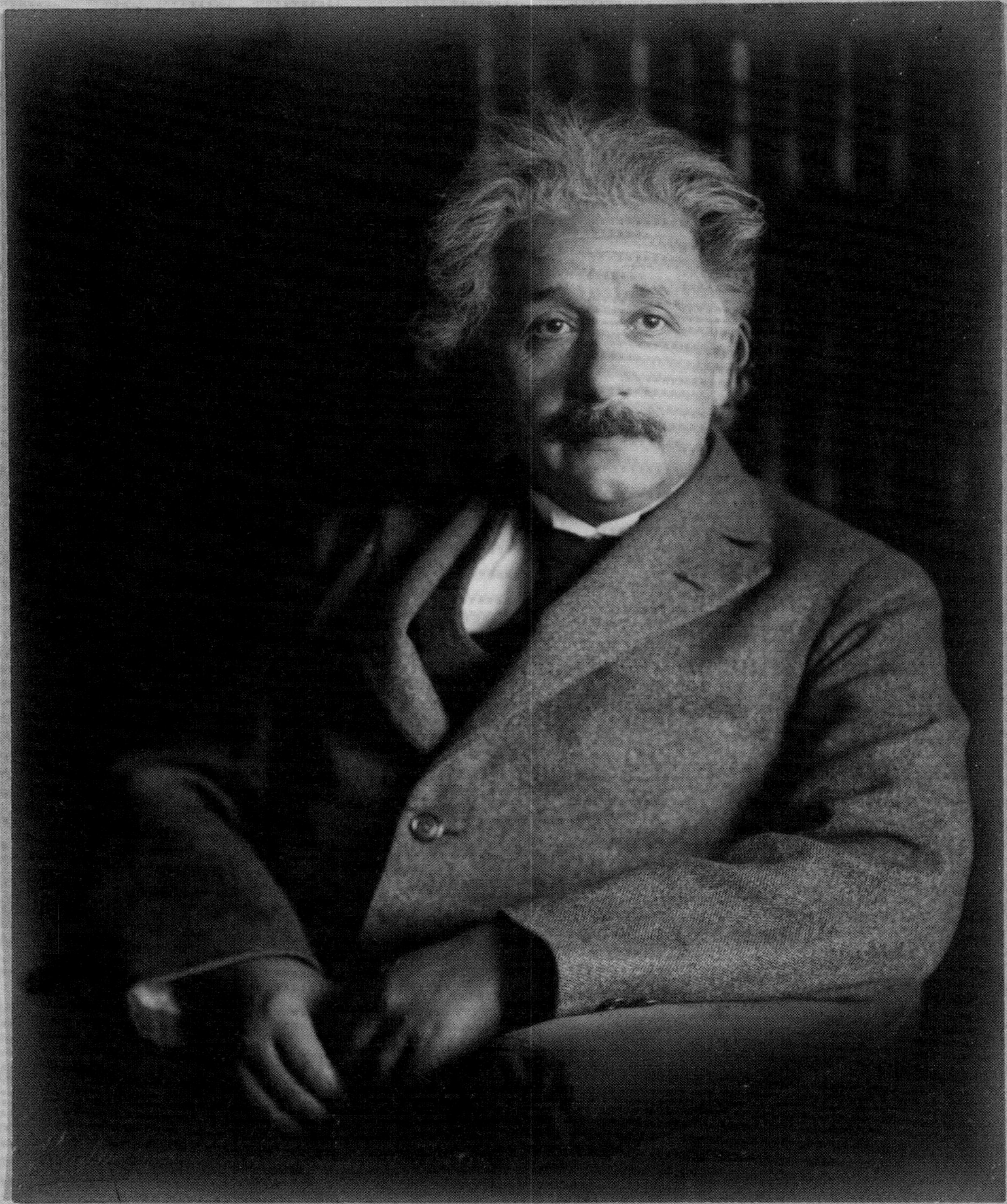

Meinem lieben Freund Plesch
zum 50.
Albert Einstein 1928.

E. Bieber
HOFPHOTOGRAPH
BERLIN

## *MY DEAR FRIEND PLESCH*

In this beautiful photograph, Einstein sits and looks straight into the camera. He presented the photo to his personal physician and good friend, Janos Plesch, for Plesch's fiftieth birthday. When this photograph was taken, Einstein was staying at Plesch's country estate outside of Berlin, recovering from an inflammatory heart condition.

"Einstein first met Plesch in 1919, when Plesch was caring for Einstein's mother, Pauline, during her terminal illness. Einstein chose Plesch to be his own physician when he became gravely ill with heart problems in 1928. Plesch guided him to recovery, part of which took place at Plesch's palatial villa in Gatow, where Einstein enjoyed privacy in the lakes region of southwestern Berlin. The two men enjoyed each other's ribald sense of humor. Einstein is reputed to have said 'Plesch is a pig, but he is my friend.'"

The photograph was taken at the renowned Emil Bieber studio in Berlin. The studio's blind stamp is on the bottom right of the photographer's mount. Also on the photographer's mount, Einstein's inscription reads, as translated from German: "My dear friend Plesch / on 50 / Albert Einstein 1928."

For Einstein's fiftieth birthday four months later, in 1929, Plesch arranged for the city of Berlin to grant Einstein, "its most famous citizen . . . lifelong rights to live in a country house that was part of a large lakeside estate that the city had acquired. There he would be able to escape, sail his wooden boat, and scribble his equations in serenity. It was a generous and gracious gesture. It was also a welcome one. . . . He was thrilled to accept."

Einstein and Plesch remained lifelong friends. Plesch visited Einstein in Princeton just a few weeks before Einstein's death. During that visit, he gave the ailing Einstein a box of fine cigars. Plesch recalled, "When Einstein saw them, he smiled and said, 'My God, I'll have to smoke these fast in order to enjoy them all!'"

PHOTOGRAPH BY EMIL BIEBER, *Berlin, Germany, 1928. Signed, dated, and inscribed by Einstein to Janos Plesch: "My dear friend Plesch / on 50 / Albert Einstein 1928."*

## *UNITING TWO PILLARS OF MODERN PHYSICS*

On January 30, 1929, Einstein published his most highly anticipated paper, *On the Unified Field Theory*. During the preceding month, newspapers around the world promoted it as his most ambitious work. Its publication caused a sensation.

Einstein had supposedly done the impossible, uniting the two pillars of modern physics: general relativity and quantum theory. The publication was a worldwide event, discussed and debated by scientists, philosophers, theologians, and even the general public.

"Before the paper was published, Einstein wrote a revealing letter to his good friend, Michele Besso, dated January 5, 1929. In it, he remarked upon the completion of his new theory. Translated, it reads, in part: [T]he very best thing, on which I have worked for days and half the nights, speculating and making calculations, is now completed and lying in front of me, compressed into seven pages with the title 'Unified Field Theory.' It looks antiquated, and the dear colleagues, including you, my dear, will initially stick their tongues out as far as possible. After all, these equations do not contain Planck's constant *h*. But once they have clearly reached the performance limit of the statistics craze, people will remorsefully return to the time-space concept, and then these equations will constitute a point of departure."

Excitement about the paper's implications—both Einstein's and the public's—was premature. The theory as published was soon disproven. Despite this setback, Einstein continued searching for the rest of his life for a theory that would unify general relativity and quantum mechanics.

*Pages 69–72:* CORRESPONDENCE,
*Einstein to Michele Besso, January 5, 1929.*

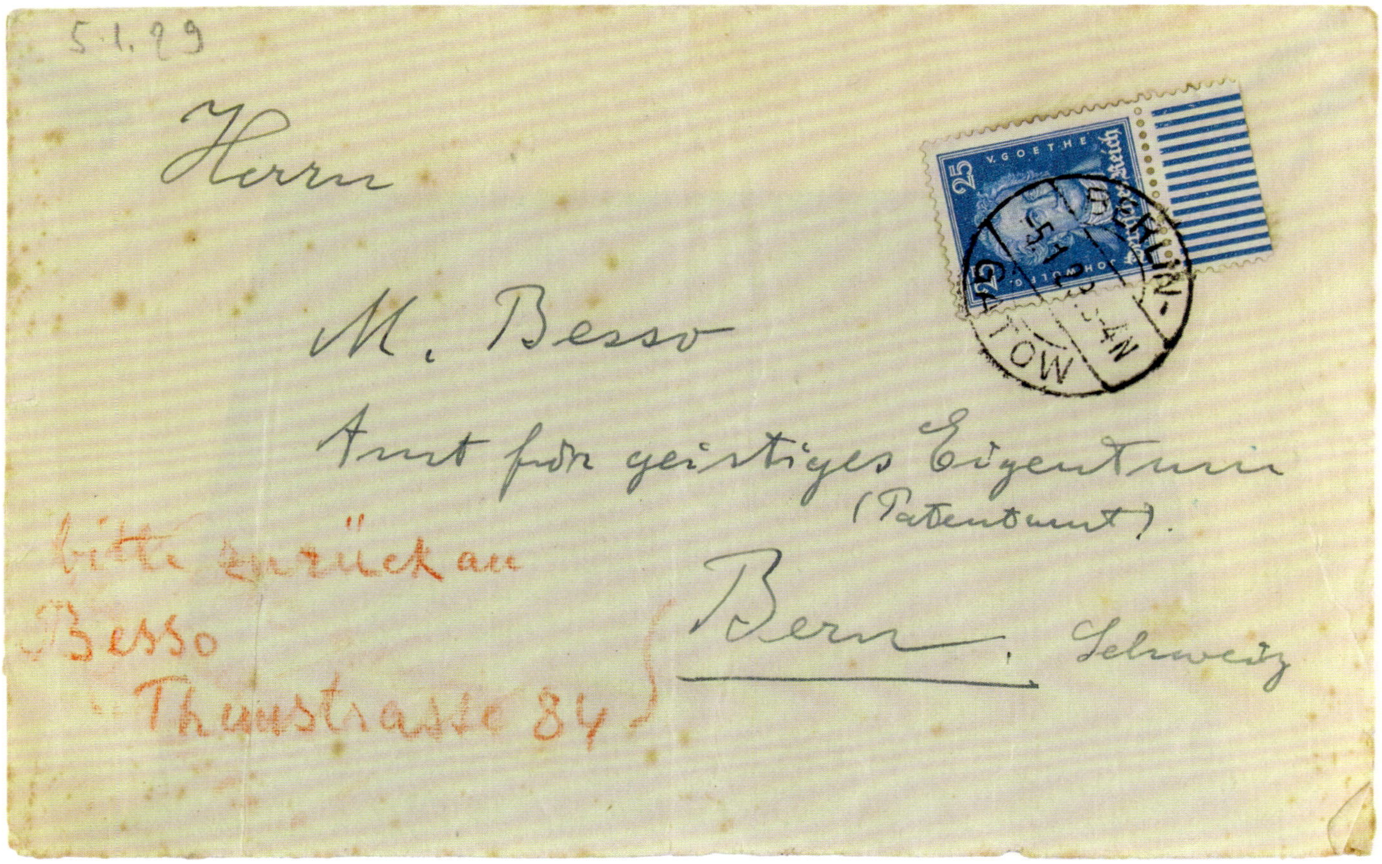
5.1.29
Herrn
M. Besso
Amt für geistiges Eigentum
(Patentamt).
Bern. Schweiz
bitte zurück an
Besso
Thunstrasse 84
V. GOETHE
Deutsches Reich

Gatow b. Berlin.

Lieber Michele!

Dank Dir für den poetischen Brief, der mich wirklich sehr freute. Während die meisten mit steigenden Jahren immer mürrischer zu werden pflegen, ist es bei Dir gerade umgekehrt. Das bewundere ich und freue mich darüber. Ich sitze nun von Zeit zu Zeit einige Wochen auf einem ländlichen Gut ganz allein in einer Wohnung und koche mir selber – wie die alten Eremiten. Dabei merkt man zu seiner Verwunderung, wie schön lange der Tag ist und wie überflüssig ein grosser Teil des geschäftigen und müssiggängerischen Treibens, worin man die übrige Zeit eingespannt ist. Ich lese mit sehr viel Spannung und Vergnügen ein Buch über Sozialismus von B. Shaw, ein ganz prächtiger Kerl mit einem grossen Einblick in das menschliche Treiben. Ich will suchen, etwas Reklame dafür zu machen. Aber das Beste, an was ich fast die ganzen Tage und die halben Nächte gegrübelt

Grüsse an Sauter und [illegible], an die anderen auch, wenn eine kleine [illegible] dabei ist.

und gerechnet habe, ist nun fertig vor mir und auf 7 Seiten zusammengepresst unter dem Namen „einheitliche Feldtheorie". Das sieht altertümlich aus und die lieben Kollegen sowie auch Du mein Lieber werden zunächst einmal die Zunge herausstrecken, so lange es geht. Denn in diesen Gleichungen kommt kein Planck'sches h vor. Aber wenn man an die Leistungsgrenzen des statistischen Fimmels deutlich gelangt sein wird, wird man wieder zur zeiträumlichen Auffassung reuevoll zurückkehren, und dann werden diese Gleichungen einen Ausgangspunkt bilden. Ich habe nämlich eine Geometrie gefunden, die nicht nur eine Riemann-Metrik sondern auch einen Fern-Parallelismus hat, den wir bisher gefühlsmässig als charakteristisch für Euklid ansahen, und die einfachsten Feldgleichungen einer solchen Mannigfaltigkeit führen zu den bekannten Gesetzen von Elektrizität und Gravitation. Selbst die Gleichungen $R_{ik} = 0$ müssen in die Rumpelkammer wandern trotz der Erfolge. Ich will nicht versäumen, Dir die Abhandlungen zu senden. Wenn Du dann Deine Zunge nicht herausstreckst, bist Du ein Heuchler; ich kenne Dich, mein Junge, sagt der Berliner.

Dir und Anna herzlichste Grüsse und Wünsche für 29.

Dein A. E.

Mit der Gesundheit geht es langsam besser. Aber ich war nahe am Abkratzen, das man ja auch nicht ungebührlich hinausschieben soll.

er mich wirklich sehr freute. Währ
de meisten mit steigenden Jahren
mürrischer zu werden pflegen, ist es
r gerade umgekehrt. Das bewunde
ls und freue mich darüber. Ich
nur von Zeit zu Zeit einige Woche
f einem ländlichen Gut ganz allein
ner Wohnung und koche mir selb
wie die alten Eremiten. Dabei
men zu seiner Verwunderung, wie se
enige der Tag ist und wie überflüs
in grosser Teil des geschäftigen und
müssiggängerischen Treibens, worin
die übrige Zeit eingespannt ist. Ich
ndet sehr viel Spannung und Vergnü
s Buch über Sozialismus von B. S
ein ganz prächtiger Kerl mit einem g
Einblick in das menschliche Treiben
will suchen, etwas Reklame dafür zu ma

*“[T]he very best thing, on which I have worked for days and half the nights, speculating and making calculations, is now completed and lying in front of me, compressed into seven pages with the title ‘Unified Field Theory.’ It looks antiquated, and the dear colleagues, including you, my dear, will initially stick their tongues out as far as possible. After all, these equations do not contain Planck’s constant* h. *But once they have clearly reached the performance limit of the statistics craze, people will remorsefully return to the time-space concept, and then these equations will constitute a point of departure.”*

—ALBERT EINSTEIN, *1929*

## *AT FIFTY*

On March 14, 1929, Einstein turned fifty years old. The tremendous outpouring of birthday wishes he received—from young and old, dignitaries and students—underscored his mass appeal.

This photo, showing a fashionably dressed and stately Einstein, was taken by the German Jewish photographer, Gerty Simon. Simon ran a successful studio in Berlin before fleeing the Nazis, ultimately settling in England.

Einstein inscribed the photo to Samuel von Fischer, founder in 1886 of the publishing house S. Fischer Verlag. Fischer was influential in the German literary scene, publishing literary giants Franz Kafka, Thomas Mann, and Hermann Hesse. S. Fischer Verlag remains one of the most prestigious publishing houses in Europe.

The inscription, translated from the German, reads: "S. Fischer, the successful sponsor of literary writings / for his seventieth birthday. / Albert Einstein. 1929."

PHOTOGRAPH BY GERTY SIMON, *Germany, March 14, 1929. Signed and dated by Einstein and signed by the photographer. Inscribed to Samuel von Fischer: "S. Fischer, the successful sponsor of literary writings / for his seventieth birthday. / Albert Einstein. 1929."*

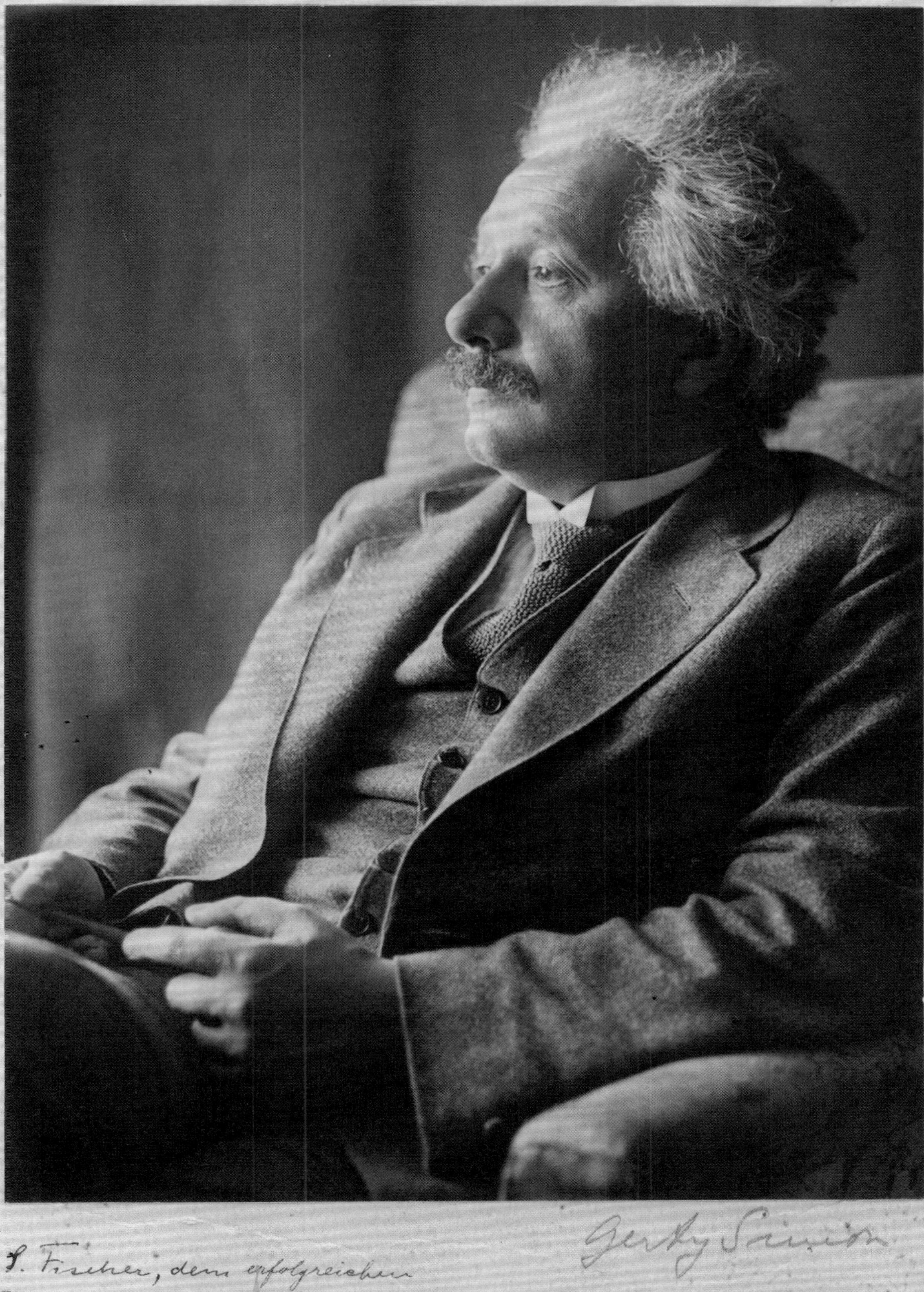

Gerty Simon

S. Fischer, dem erfolgreichen
Förderer literarischen Schaffens
zum siebzigsten Geburtstage
Albert Einstein. 1929.

## *SCIENCE AND ART*

We get a rare glimpse of Einstein's mind at work in this autograph manuscript from around 1929. Here, he was working on problems related to his life's work: understanding gravitational forces, quantum theory, and the implications of his general theory of relativity.

His words and equations are complementary. Following the first series of equations, he writes, in German, "Gravity effect on particles neglected: not possible, because gravity field generating effect of particles neglected. A more promising approach would be via equations."

Einstein claimed that "the greatest scientists are artists as well." This work attests to that notion—Einstein, the scientific artist—demonstrating the unique craft of his explorations of the universe.

*Pages 77–79:* ALBERT EINSTEIN,
*autographed manuscript page, ca. 1929.*

| $h$ | $\alpha_2$ | $\varphi$ | $\gamma$ | $\alpha_3$ und $\alpha_4$ | |
|---|---|---|---|---|---|
| 2 | 4 | 0 | 1 | 2 | 2 |

$\alpha_3 \quad g_{\iota\kappa,\sigma} \quad g_{\lambda\mu,\sigma} \quad g^{\iota\lambda} \quad g^{\kappa\mu}$

$\underset{2}{\alpha_3} \quad \underset{1}{\gamma_{\iota\kappa,\nu}} \quad \underset{1}{\gamma_{\lambda\mu,\sigma}} \quad \eta_{\iota\lambda} \quad \eta_{\kappa\mu}$

Gravitationswirkung auf Teilchen vernachlässigt.

geht nicht, weil ~~felderzeugende~~ gekürt Wirkung von Teilchen vernachlässigt.

Für grav. Näherung muss von Gleichungen ausgegangen werden.

| $h_2$ | $\alpha_2$ | $\varphi$ | $\gamma$ | $\alpha_3$ | $\alpha_4$ |
|---|---|---|---|---|---|
| 2 | 2 | 0 | 1 | 0 | 0 |

Für die $\gamma$ und $\varphi$ 17 Gleichungen

Ausserdem 4 Gleichungen durch Elimination der $h$ aus den Gravitationsgleichungen

Ist nur möglich, wenn 4 Identitäten bestehen.

---

$$\mathfrak{E} = \int T_{44}\, dV = h\nu\,\mathfrak{E}$$

$$\frac{\varepsilon^2}{l} = h\nu \quad \Big| \quad \frac{\varepsilon^2}{c} = h \quad \Big| \quad \varepsilon^2 \sim h(F)$$

$$h \quad \Big| \quad \frac{\int T_{44}\, dV}{\nu} \quad \Big| \quad \frac{\varepsilon^2}{(F)}$$

$\varepsilon$ Oberfl. Integr. Wenn $\int T_{44}\, dV$ in Oberfl. Integr. verwandelbar, dann ~~h~~ berechenbar. Dies Integral müsste auf $\varepsilon$ zurückführbar sein, da dies das einzige berechenbare Volumintegral sein dürfte.

| $h$ | $\alpha_2$ | $\varphi$ | $\gamma$ | $\alpha_3$ und $\alpha_4$ |
|---|---|---|---|---|
| 2 | ~~4~~ 2 | 0 | 1 | 2 · 2 |

Gravitationswirkung auf Teilchen verr
geht nicht, weil ~~felderzeugende~~ gewirkt Wirk
Für grav. Näherung muss von Gleich

| $h_2$ | $\alpha_2$ | $\varphi$ | $\gamma$ | $\alpha_3$ | $\alpha_4$ |
|---|---|---|---|---|---|
| 2 | 2 | 0 | 1 | 0 | 0 |

Für die $\gamma$ und $\varphi$ 14 Gleichung

Ausserdem 4 Gleichungen durch Ele

Ist nur möglich, wenn 4 Iden

---

$$\mathfrak{E} = \int T_{44}\, dV = h\nu \xi$$

$\int$

$\alpha_3 \; g_{\iota\kappa,0} \; g_{\lambda\mu,0} \; g^{\iota\lambda} \; g^{\kappa\mu}$

$\underset{2}{\alpha_3} \; \underset{1}{\gamma_{\iota\kappa,0}} \; \underset{1}{\gamma_{\lambda\mu,0}} \; \eta_{\iota\lambda} \; \eta_{\kappa\mu}$

...achlässigt.

...ng von Teilchen vernachlässigt.

...ungen ausgegangen werden.

...

...mination der h aus den Gravitationsg...

...itäten bestehen.

---

$\frac{\varepsilon^2}{l} = h\nu \;\Big|\; \frac{\varepsilon^2}{c} = h \;\Big|\; \varepsilon^2 \sim$

## *LISTENING TO MUSIC*

This candid image shows Einstein in his study, absorbed by music and perhaps engaging in some private conducting.

Music was so important to Einstein that, in a 1929 interview with George Sylvester Viereck for the *Saturday Evening Post*, he said, "If I were not a physicist, I would probably be a musician. I often think in music. I live my daydreams in music. I see my life in terms of music." In other interviews, Einstein associated music with his scientific insights, intuition, and creativity. His son Hans Albert reported that "whenever he felt that he had come to the end of the road or into a difficult situation in his work, he would take refuge in music, and that would usually resolve all his difficulties."

Einstein inscribed the photograph to a Miss Stires for her birthday, dated January 31, 1930.

UNKNOWN PHOTOGRAPHER, *1930.*
*Signed, dated, and inscribed by Einstein to a Miss Stires.*

*If I were not a physicist,*
*I would probably be a musician.*
*I often think in music.*
*I live my daydreams in music.*
*I see my life in terms of music."*

—ALBERT EINSTEIN, *1929*

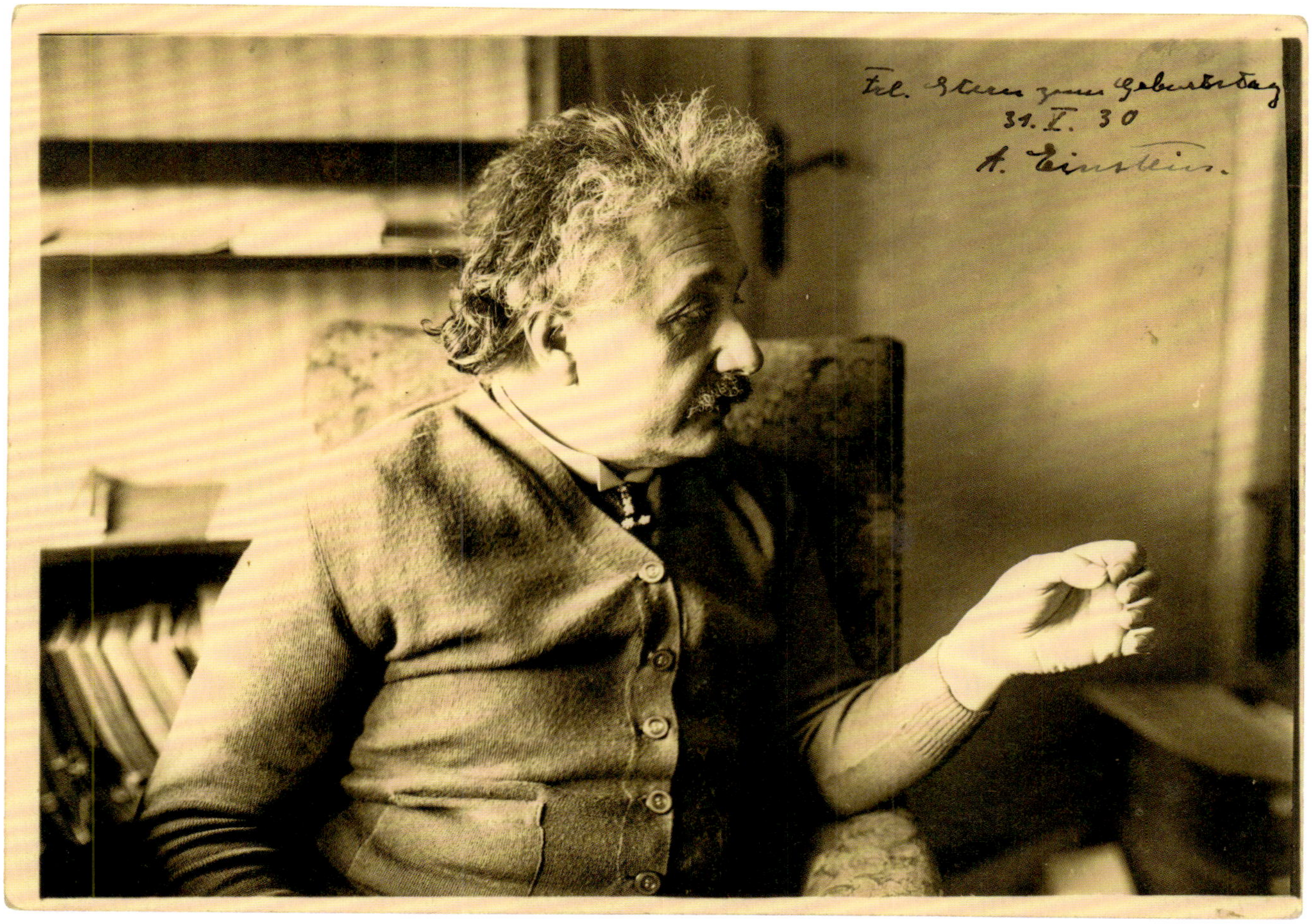
Frl. Stern zum Geburtstag
31. V. 30
A. Einstein.

### *SCIENCE, REALITY, MATTER*

In 1930, Dr. Ralph Lyndal Worrall of the Royal Society of Medicine was working on his book *The Outlook of Science: Modern Materialism*. For this work, he sought opinions about scientific materialism from the greatest minds of the day. Naturally, Worrall wrote to Einstein for his thoughts on the subject.

Responding to Worrall's questions about the relationship between science, reality, and matter, Einstein replied in this letter, translated from the original German: "All physics is realistic insofar as it starts from the hypothesis of a reality that is independent of perception and thought. It is no longer materialistic inasmuch as it does not acknowledge matter as the irreducible conceptual basis of the physical theoretical system."

Here, Einstein identifies realism as the philosophical foundation of physics but asserts that the basis of the "real" is not matter. Einstein believed that "matter" itself is an invention of the mind in its encounter with nature or reality, and matter is merely one possible construct.

In asserting that matter is no longer "the irreducible conceptual basis" of physics, Einstein signaled that theoretical physics had moved to a more fundamental concept, which his work had fostered when he showed that matter and energy are different expressions of the same thing.

CORRESPONDENCE, *Einstein to Dr. Ralph Lyndal Worrall of the Royal Society of Medicine, March 18, 1930.*

ALBERT EINSTEIN

BERLIN W, den 18. März 1930
HABERLANDSTR. 5

Herrn Dr. R.L. Worrall
Royal Society of Medicine
1 Wimpole St.
London W.I

Sehr geehrter Herr!

Soviel ich bei oberflächlicher Einsicht habe sehen können, stimme ich mit Ihnen überein. Alle Physik ist insofern realistisc als sie von der Hypothese einer von der Wahrnehmung und vom Denken unabhängigen Wirklichkeit ausgeht. Sie ist nur insofern nicht mehr materialistisch , als sie die Materie nicht ~~xxxx~~ als irreduzible begriffliche Basis des physikalischen theoretischen Systems anerkennt.

Mit vorzüglicher Hochachtung

A. Einstein

## *A MILITANT PACIFIST*

Einstein spent most of 1930 in Germany before leaving in December for his second visit to the United States. The Nazi party was gaining strength in the government. Einstein recognized the danger and became increasingly outspoken about his lifelong commitment to pacifism. He told one reporter, "I am not only a pacifist, I am a militant pacifist."

This photograph is from Studio Stone, run by the Jewish husband and wife team of Sasha and Cami Stone in Berlin from 1924 to 1931. Operating a successful photography studio for commercial work, Cami and Sasha Stone created a series of highly regarded photographs for German avant-garde publications. In 1931 the Stones were able to leave Germany and move to Belgium, where Cami was born and had maintained citizenship.

This stunning photograph captures a middle-aged Einstein in a contemplative pose, looking downward with his brow slightly furrowed. It was likely taken in 1929 or 1930 by Cami Stone. The circumstances of his signing and dating it later are unknown.

PHOTOGRAPH BY STUDIO STONE
*(Sasha and Cami Stone), Berlin, Germany, ca. 1930. Signed and dated by Einstein in 1932.*

*No person has the moral right to call himself a Christian or Jew so long as he is prepared to engage in systematic murder at the command of an authority, or allow himself to be used in any way in the service of war or the preparation for it."*

—ALBERT EINSTEIN, *ca. 1930*

"No person has the moral right to call himself a Christian or Jew so long as he is prepared to engage in systematic murder at the command of an authority, or allow himself to be used in any way in the service of war or the preparation for it."

—ALBERT EINSTEIN, ca. 1930

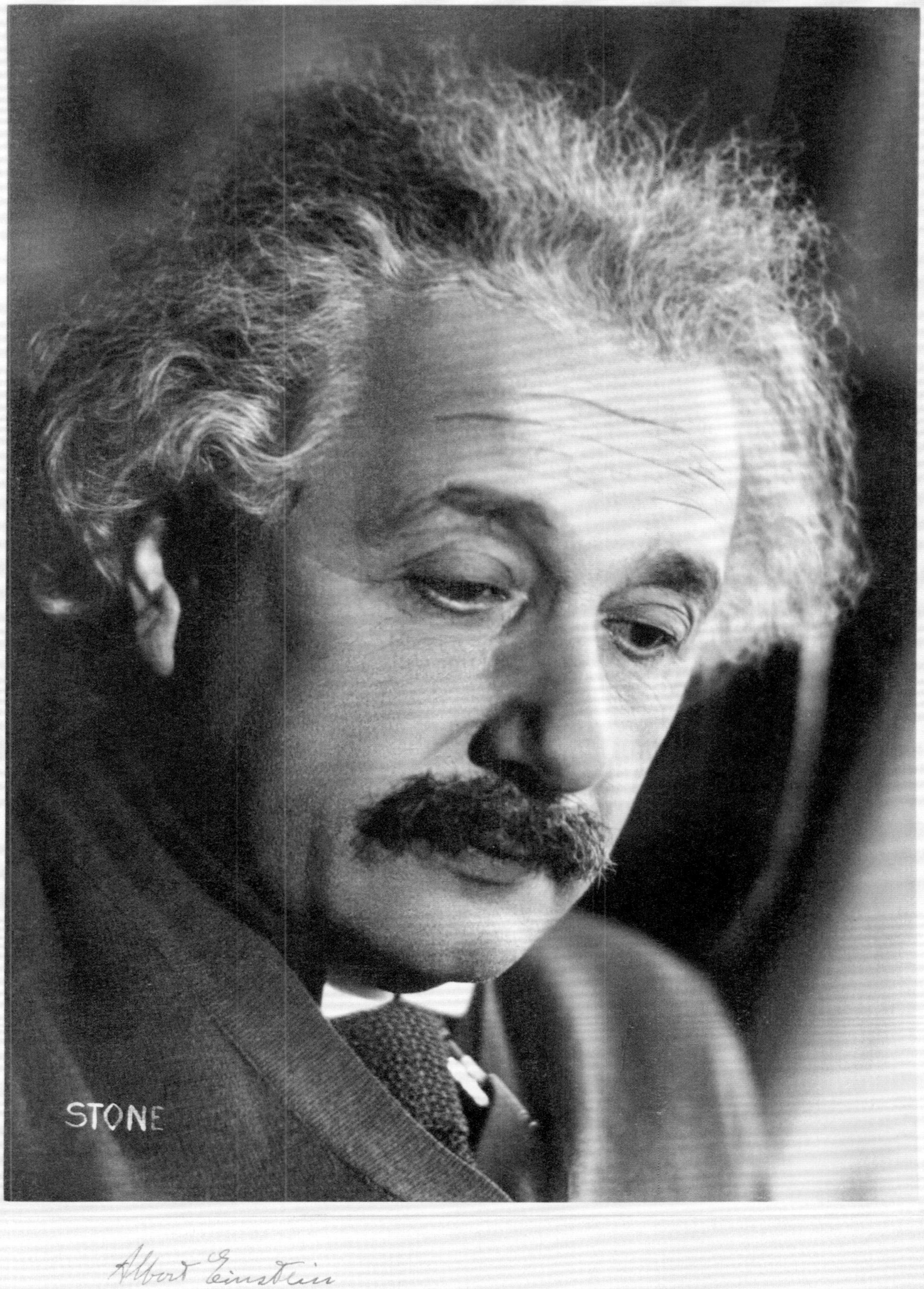


Albert Einstein
1932.

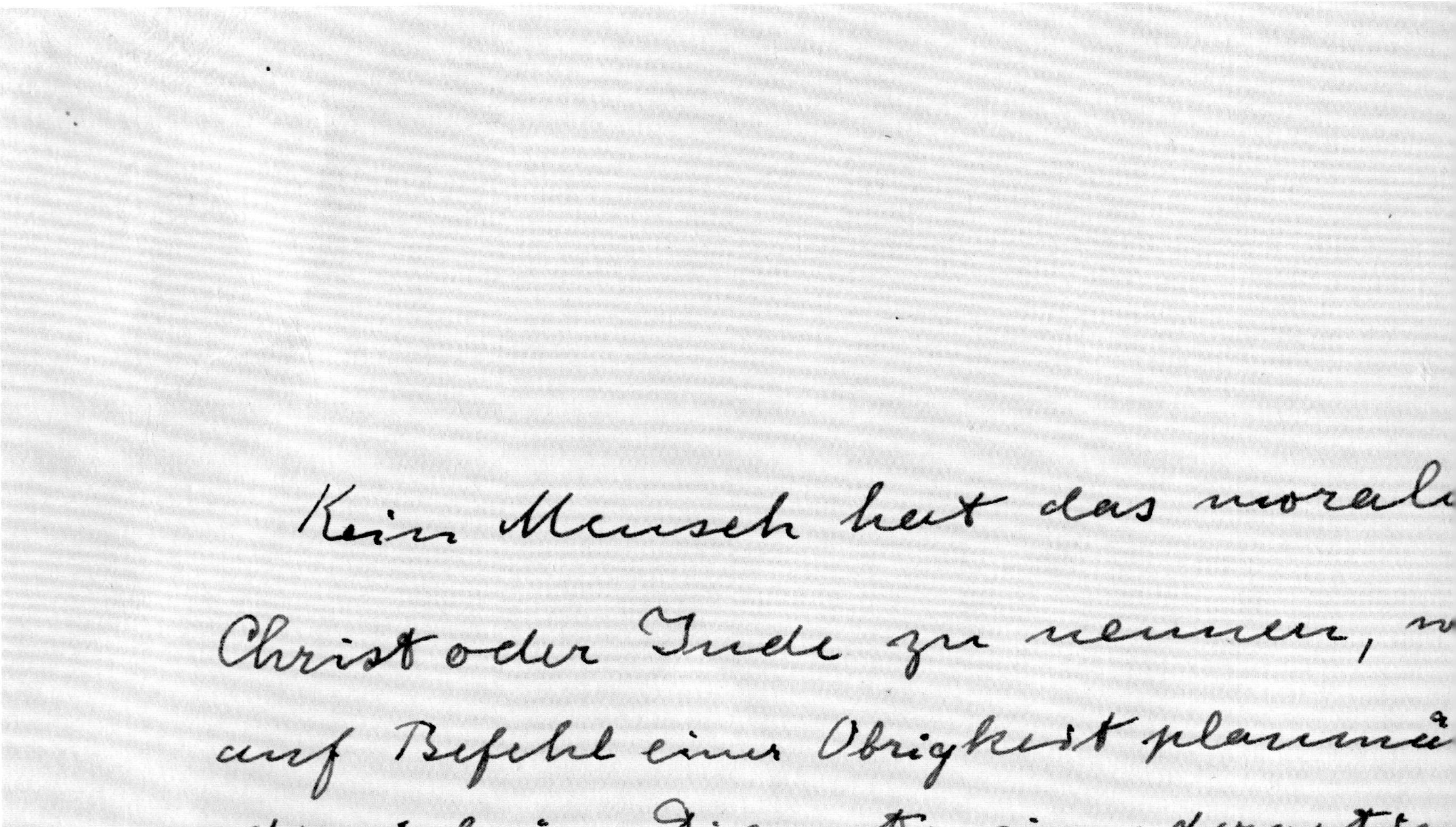
Kein Mensch hat das morali
Christ oder Jude zu nennen, w
auf Befehl einer Obrigkeit planmä
oder sich im Dienste eines derartig
oder der Vorbereitungen hiefür irge
zu lassen

A. Einstein.

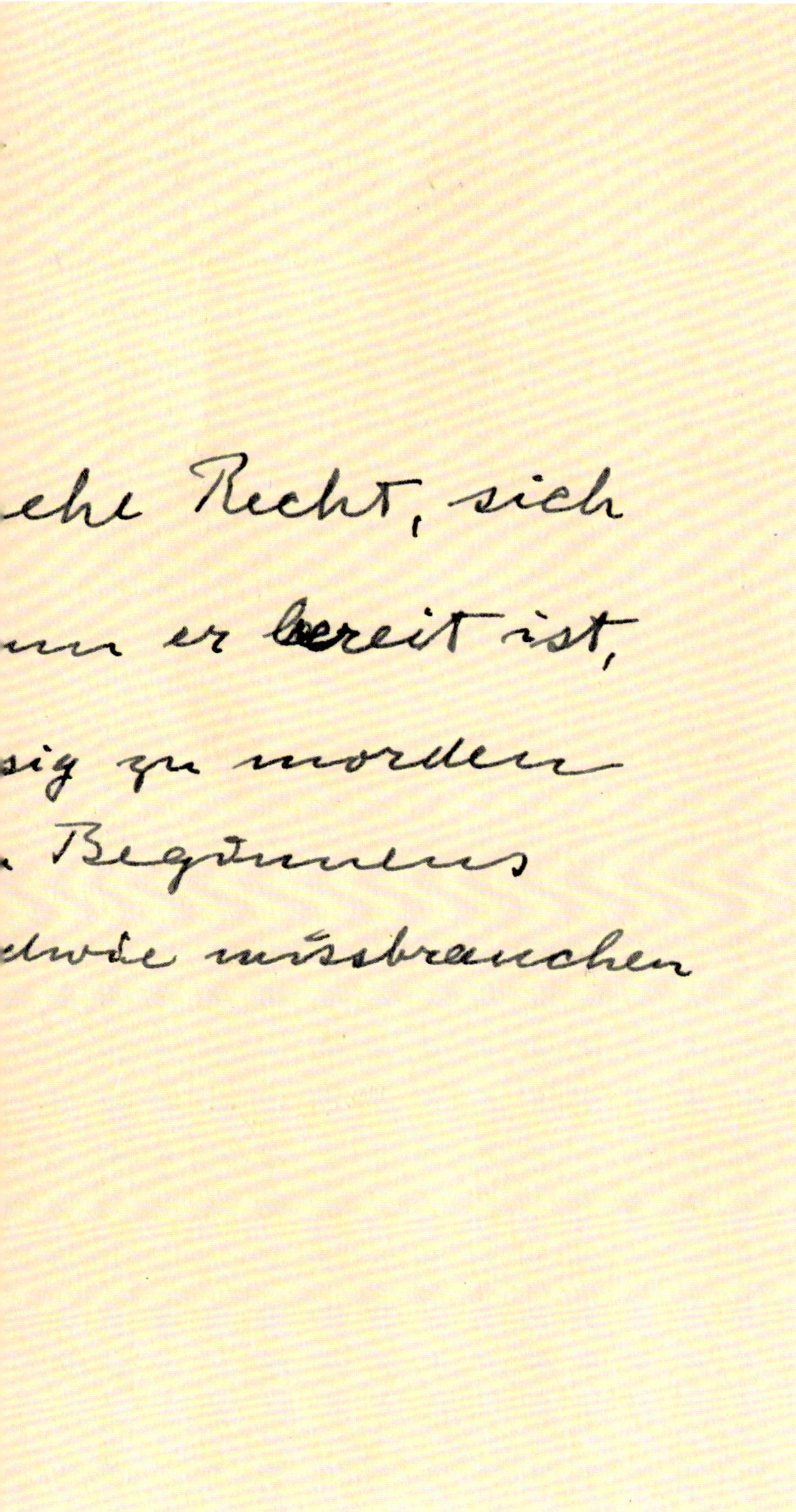


***PAX MUNDI***

In 1925, the French Academy started sending large blank folio sheets to influential scientists, writers, and intellectuals, asking them to write a statement about peace. Seven years later, the World League for Peace released a limited edition of prints of the submissions under the title *Pax mundi: Livre d'or de la paix.*

Einstein responded with this handwritten expression of his lifelong dedication to pacifism. Translated from the original German, it reads:

"No person has the moral right to call himself a Christian or Jew so long as he is prepared to engage in systematic murder at the command of an authority, or allow himself to be used in any way in the service of war or the preparation for it."

ALBERT EINSTEIN, *submission to the World League for Peace publication* Pax mundi: Livre d'or de la paix (World Peace: The Golden Book of Peace), *ca. 1925–32.*

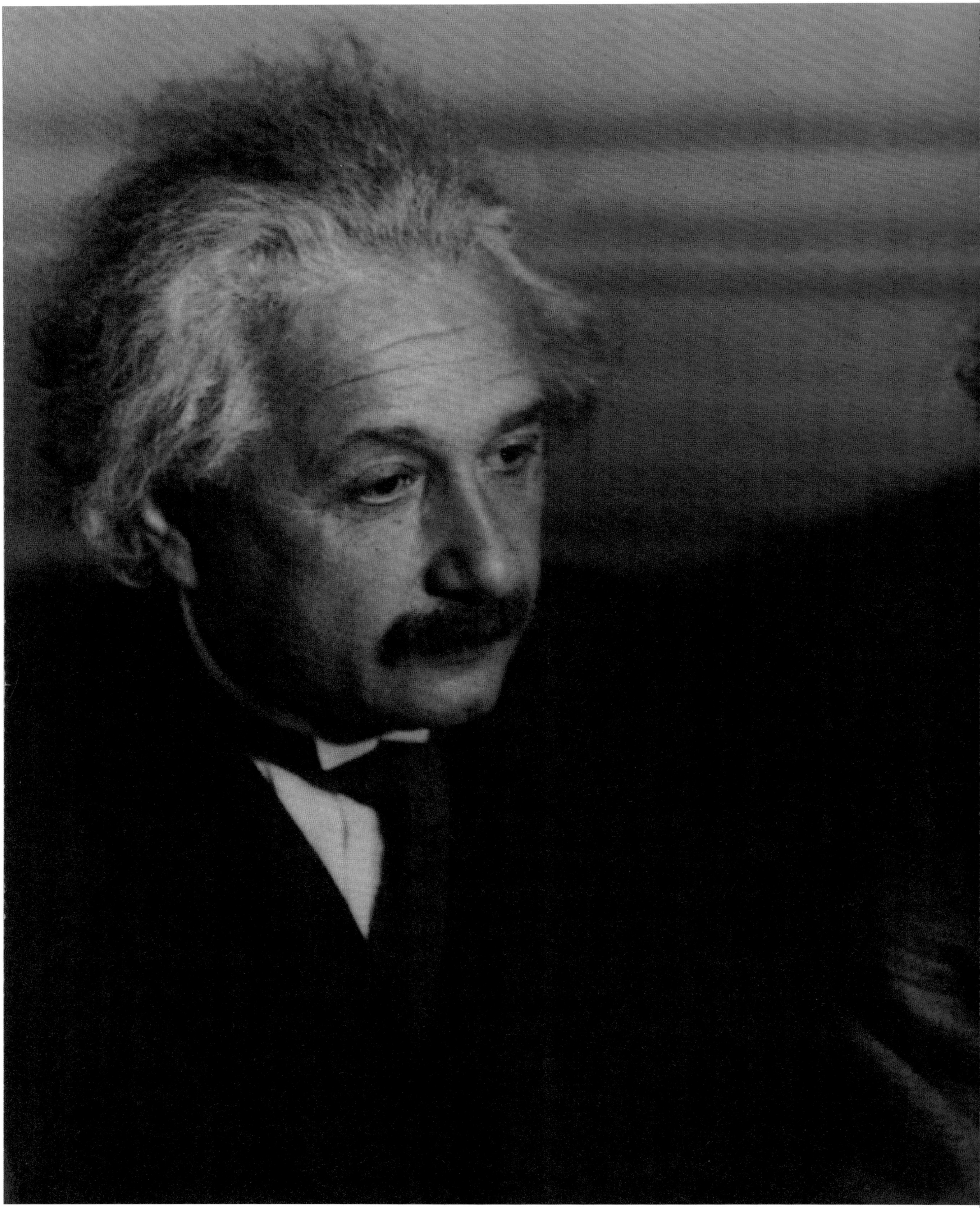

# 1930

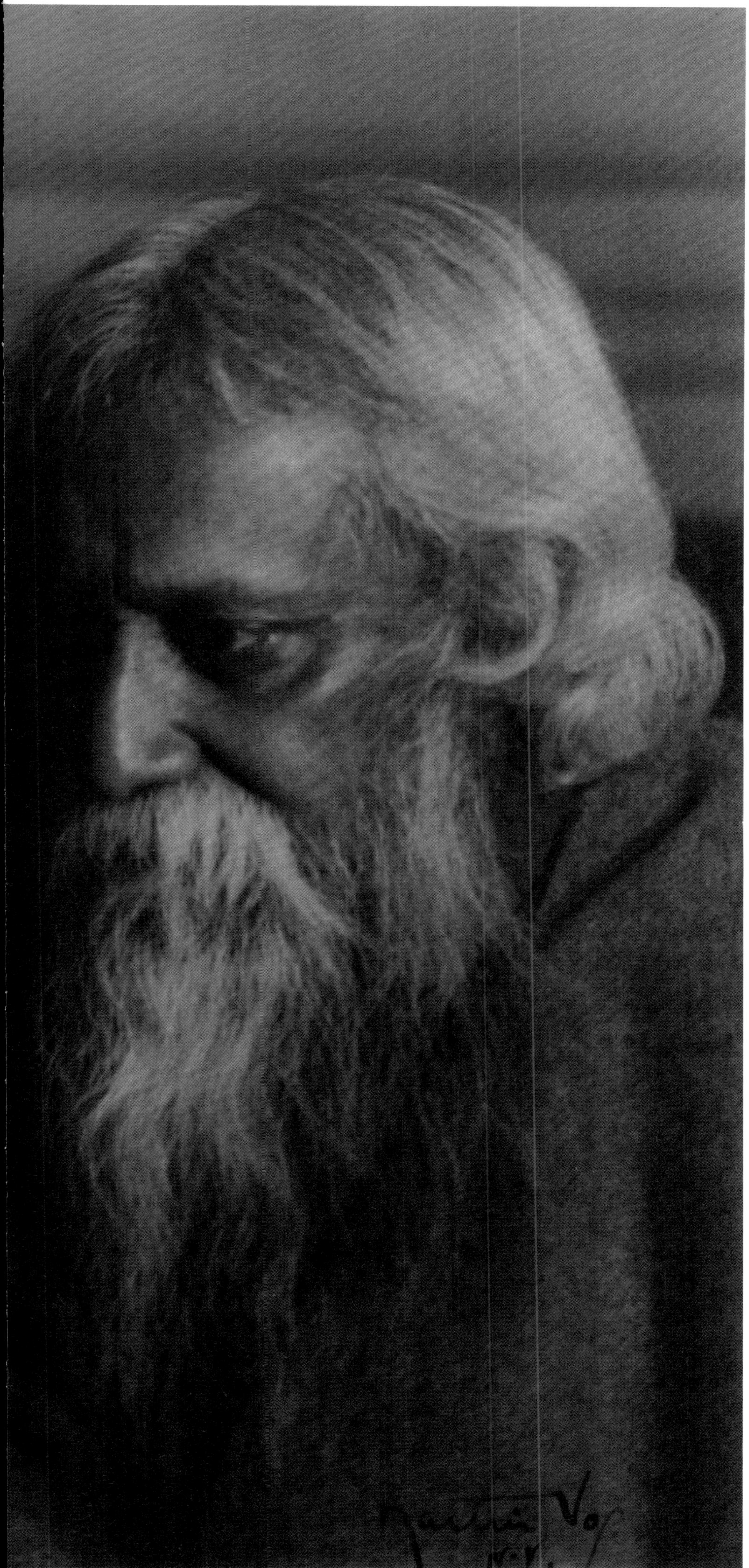

## *TWO PLANETS*

This oversized vintage photograph by American photographer Martin Vos captures a historic encounter between Nobel laureates Albert Einstein (Physics, 1921) and Rabindranath Tagore (Literature, 1913).

Einstein and Tagore met on July 14, 1930, at Einstein's lakeside estate in Caputh that had been bequeathed to him by the city of Berlin. There, the two great thinkers discussed the nature of reality.

Tagore framed their approaches as follows: "You have been busy hunting down with mathematics the two ancient entities, time and space, while I have been lecturing in this country on the eternal world of man, the universe of reality."

*The New York Times Magazine* published their conversation on August 10 under the headline "Einstein and Tagore Plumb the Truth," authored by the Russian journalist Dimitri Marianoff.

Marianoff wrote in his preamble: "It was interesting to see them together—Tagore, the poet with the head of a thinker, and Einstein, the thinker with the head of a poet. . . . Neither sought to press his opinion. But it seemed to an observer as though two planets were engaged in a chat."

PHOTOGRAPH BY MARTIN VOS,
*Caputh, Germany, 1930. Einstein with Rabindranath Tagore, winner of the Nobel Prize in Literature. Signed by the photographer.*

### *A SECOND KEY TO THE CITY*

Einstein arrived in New York City for the second time on December 11, 1930. He had hoped to avoid publicity and have a short, simple stay before leaving for California. But the American public's interest in him had only swelled during the decade since his initial visit. Even before disembarking, he was swamped by reporters and flooded with invitations. On December 13, when Einstein inscribed this photo, he received a second key to the city from New York's mayor, Jimmy Walker, in a ceremony at City Hall.

The studio portrait shows Einstein in an unusual pose, with his body in profile and his head turned toward the camera. It was almost certainly taken in Germany before his visit. The photographer is unknown.

The recipient, Bernhard Edelhertz, was the New York–based publisher, secretary, and treasurer of *The American Hebrew* magazine. Einstein's photograph presentation to an important New York Jewish figure underscored his desire to strengthen relationships with the American Jewish population. During his brief stay in New York, he attended a Zionist rally, delivered a radio address at NBC studios to Zionist youth, and celebrated Hanukkah at a large gathering at Madison Square Garden.

UNKNOWN PHOTOGRAPHER,
*Germany, ca. 1930. Signed, dated, and inscribed by Einstein to Bernhard Edelhertz.*

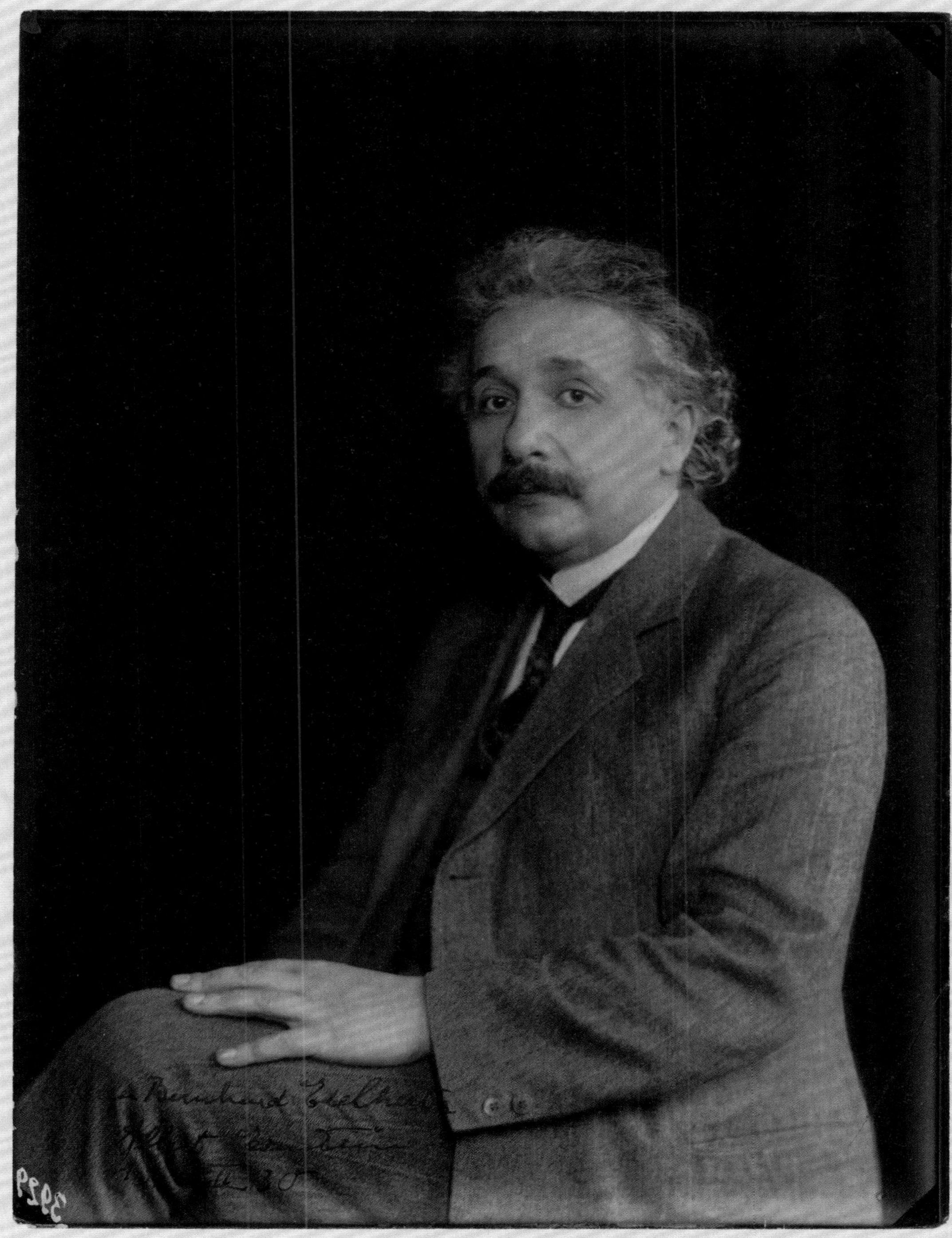

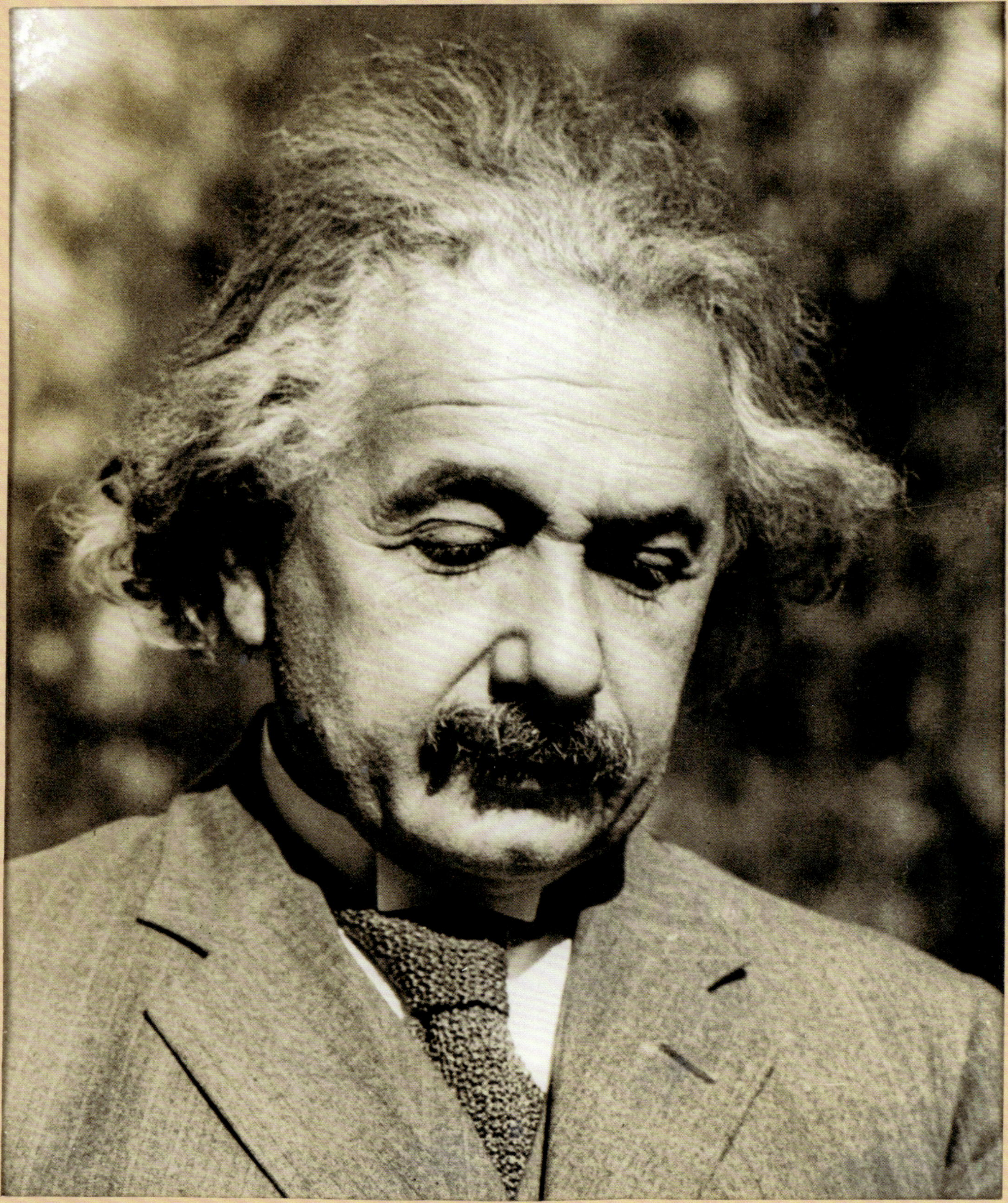

A. Einstein

### *ON THE BACK OF AN OLD ENVELOPE*

This striking photograph, showing a handsome and serious Einstein bathed in sunshine, is undated and the photographer is unknown. Our research suggests that, like the image that follows, it was taken on one of his trips to California in the early 1930s. During his second visit, in January 1931, Einstein went to see the 100-inch Hooker telescope at Mount Wilson Observatory. It was the world's largest telescope at that time, and Edwin Hubble used it to make his monumental discovery of the expanding universe.

Einstein met with Hubble and examined the famous telescope. Einstein's wife Elsa had accompanied him on this trip. When she was shown the sophisticated equipment used to explore the shape and behavior of the universe, she reportedly quipped, "Well, my husband does that on the back of an old envelope."

UNKNOWN PHOTOGRAPHER,
*ca. 1930. Signed by Einstein.*

## *CITY LIGHTS*

Einstein visited California three separate times in the early 1930s. He mainly intended to meet with scientists at the California Institute of Technology. His presence in the Golden State did not escape public notice, and his schedule quickly filled up. By 1931, even people who had little understanding of Einstein's scientific achievements idolized him. He was granted a reception usually reserved for Hollywood elites.

Einstein posed for pictures with Charlie Chaplin, purportedly the only celebrity he specifically requested to meet, at the 1931 premiere of Chaplin's film *City Lights*. The encounter produced an oft-quoted, perhaps apocryphal, exchange:

Einstein: "What I most admire about your art is your universality. You don't say a word, yet the world understands you!"

Chaplin: "True. But your glory is even greater! The whole world admires you, even though they don't understand a word of what you say."

This press photo, taken during one of his California visits, shows Einstein relaxed, smiling, and with binoculars around his neck, looking very much the tourist. He signed the print in 1953.

UNKNOWN PHOTOGRAPHER,
*California, ca. 1931. Signed and dated by Einstein in 1953.*

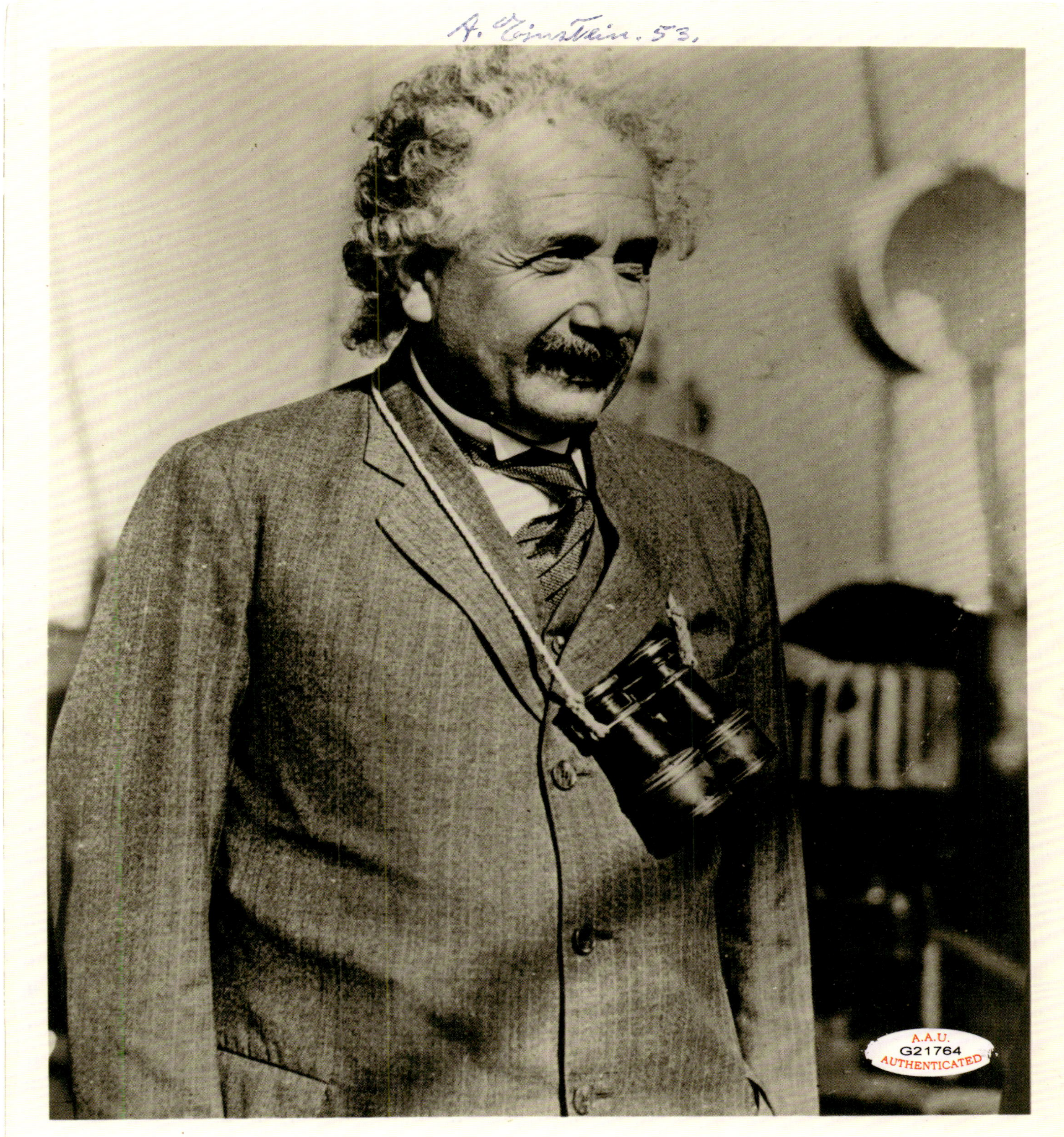
A. Einstein. 53.
A.A.U.
G21764
AUTHENTICATED

## *THE EXPANDING UNIVERSE*

When Einstein created the general theory of relativity, he believed—along with the rest of the scientific community—the universe was static. However, his field equations implied that the universe was in a constant state of flux.

Einstein assumed there must be a flaw in his equations. He added a term to correct the supposed defect so that his field equations would be consistent with an unchanging universe. In effect, the added constant counteracted the influence of gravity. He called this term the "cosmological constant." Einstein published his modification in a 1917 paper entitled *Cosmological Considerations in the General Theory of Relativity.*

A decade later, Edwin Hubble discovered galaxies beyond our own Milky Way. Hubble located particular stars (cepheid variables) that could be used to measure their distance from Earth. He observed that the farther away the galaxies were from Earth, the greater light's redshift (lengthening of wavelength). In other words, the more distant the galaxy, the faster its velocity is away from Earth. Hubble's observations showed that the universe is not static. It is expanding.

Based on the new astronomical data, Einstein rejected the cosmological constant in the 1931 publication shown here, "On the Cosmological Problem of the General Theory of Relativity." He later said that introducing the cosmological constant to his field equations was his "greatest blunder."

Since the time of Hubble's discoveries, astronomers have determined that the universe is not only expanding but also that it is expanding at an accelerating rate.

Today's scientists believe there is a repulsive "dark energy" in space causing the increasing expansion rate. Ironically, it now seems that Einstein's "greatest blunder"—the cosmological constant—may be vital to understanding this force.

ALBERT EINSTEIN, *"Zum kosmologischen Problem der allgemeinen Relativitätstheorie" ("On the Cosmological Problem of the General Theory of Relativity") in* Sitzungsberichte der Preussischen Akademie der Wissenschaften (Report from the Prussian Academy of Sciences), *1931.*

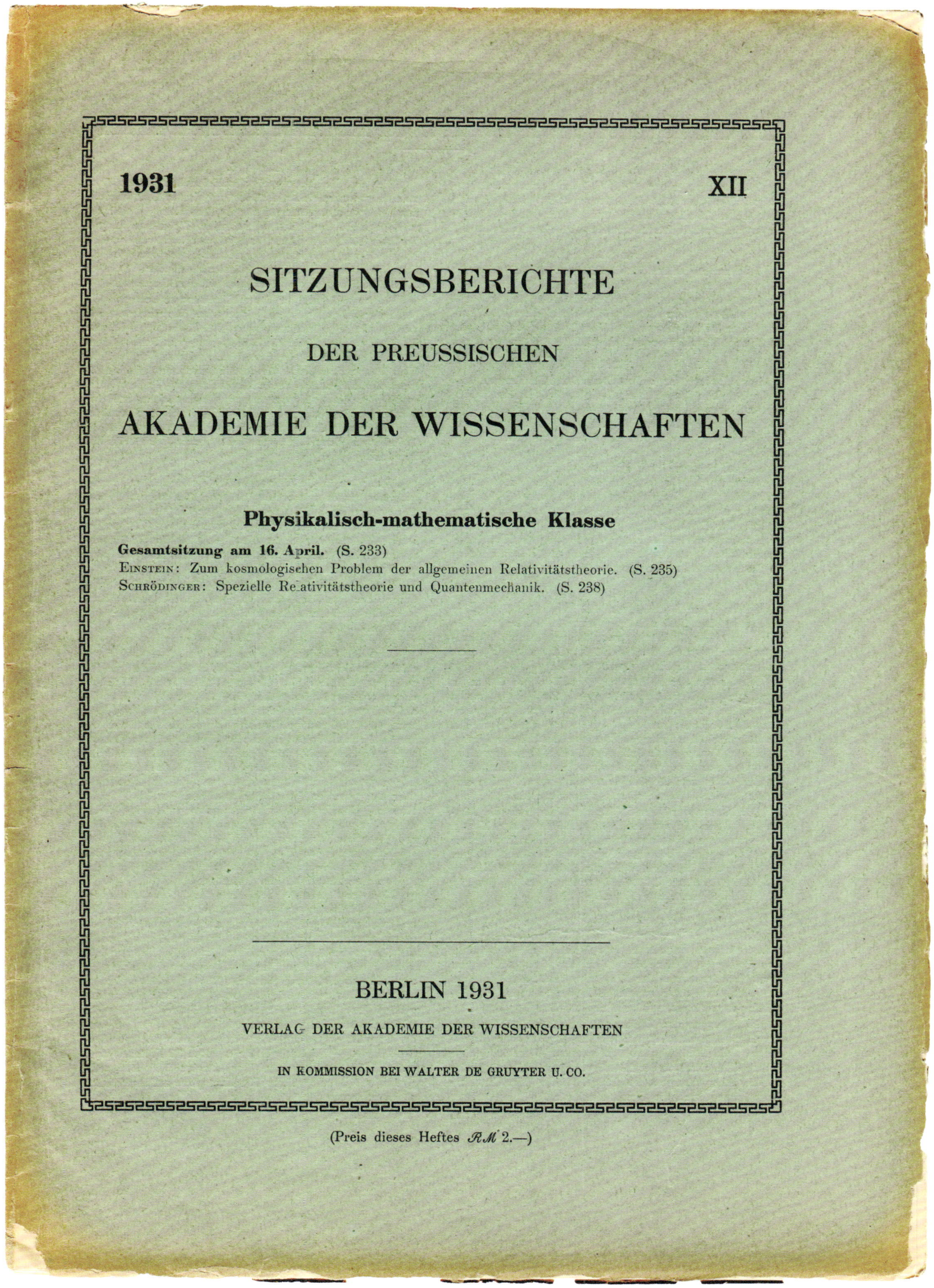

1931 XII

# SITZUNGSBERICHTE

DER PREUSSISCHEN

# AKADEMIE DER WISSENSCHAFTEN

## Physikalisch-mathematische Klasse

**Gesamtsitzung am 16. April.** (S. 233)

Einstein: Zum kosmologischen Problem der allgemeinen Relativitätstheorie. (S. 235)

Schrödinger: Spezielle Relativitätstheorie und Quantenmechanik. (S. 238)

---

BERLIN 1931

VERLAG DER AKADEMIE DER WISSENSCHAFTEN

IN KOMMISSION BEI WALTER DE GRUYTER U. CO.

(Preis dieses Heftes RM 2.—)

*"He respected children and liked their curiosity and fresh approach to life and therefore did not want to ignore them."*

—EVELYN EINSTEIN, *2002*

***RESPECT FOR CHILDREN***

This charming, candid photograph of a smiling Einstein with two young children appears to have been taken in America. Einstein's love of children is well-documented. The photo was dated and signed in full by Einstein and his wife Elsa.

UNKNOWN PHOTOGRAPHER,
*United States, 1932. Signed and dated by Einstein and signed by Elsa Einstein.*

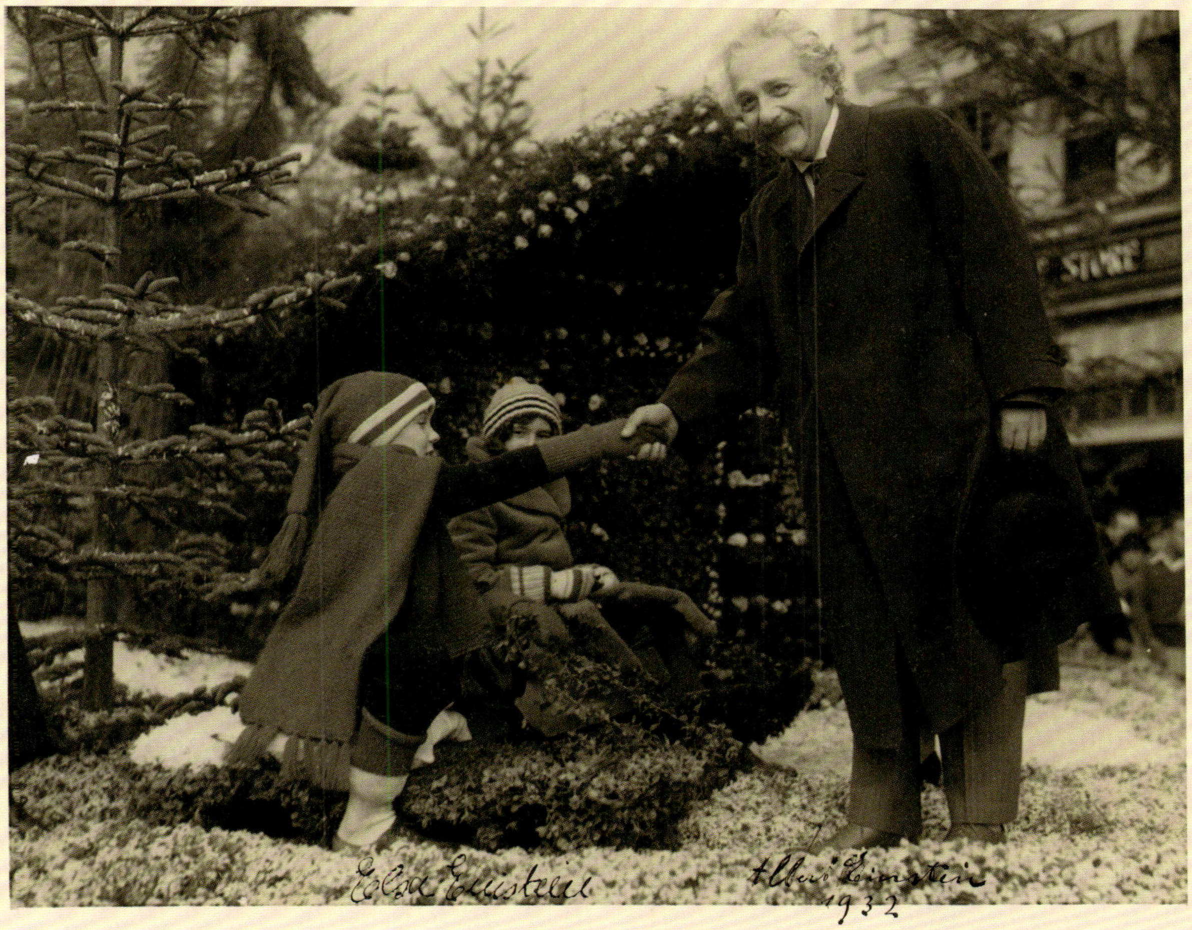
Elsa Einstein
Albert Einstein
1932

## *WORLD EVENTS INTERVENED*

Albert and Elsa Einstein arrived in California on January 9, 1933, planning a short visit before returning to Germany. World events intervened.

On January 31, Hitler assumed power as German chancellor, and the Nazis' purging of Jews began. Einstein became an enemy of the state overnight. The Nazis discredited his work, burned his publications, raided his home, and confiscated his property—even his beloved sailboat in Caputh.

Einstein resigned from all positions and affiliations in Germany. For the second time, he renounced his German citizenship. He and Elsa became refugees.

This photo was taken in Los Angeles at the Ambassador Hotel studio of noted portrait photographer Aaron Tycko. Einstein and Tycko both signed it, with Einstein's signature in full.

PHOTOGRAPH BY AARON TYCKO,
*Los Angeles, 1933. Signed and dated by Einstein.*

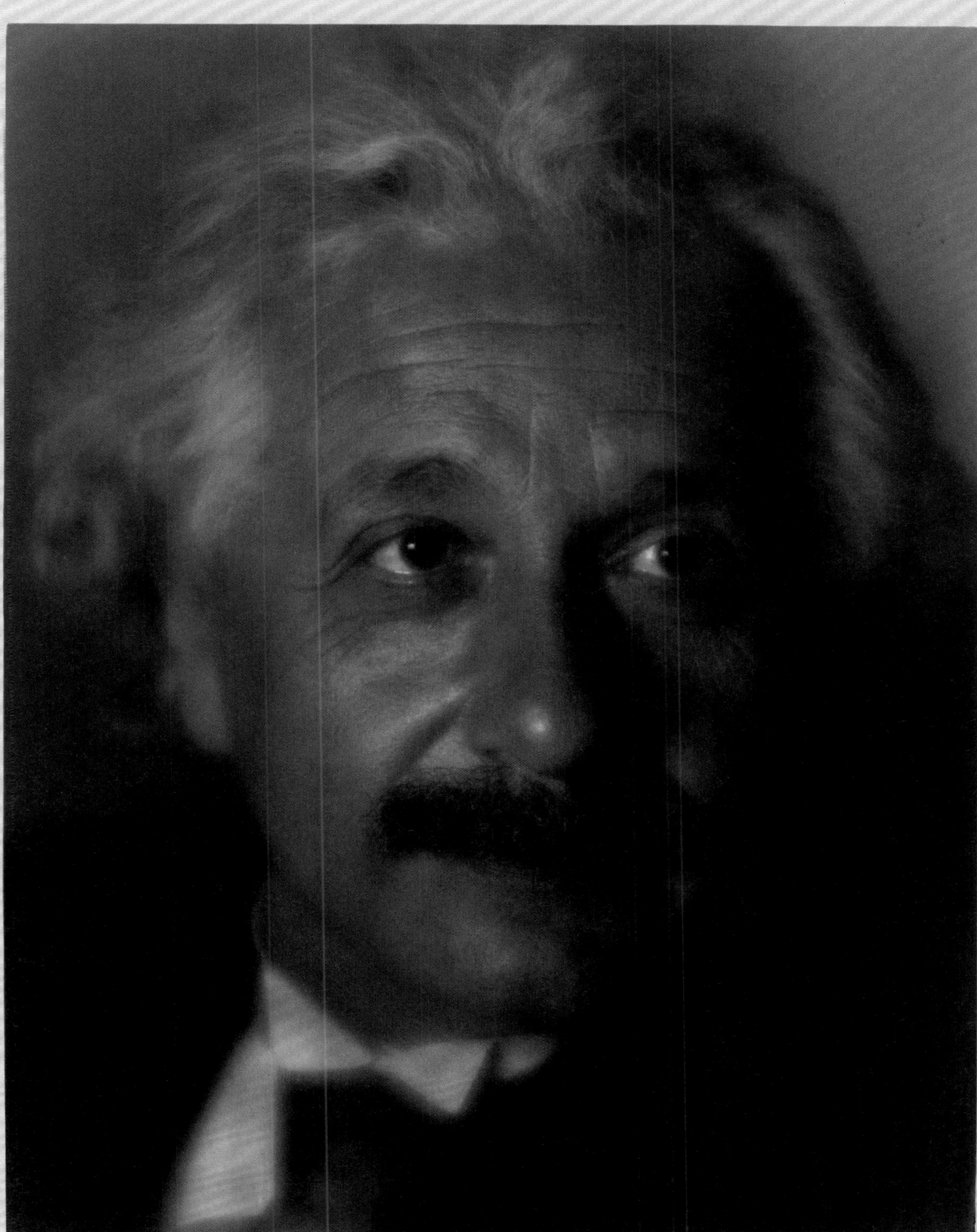

Albert Einstein 1933.

Tycko S.A.

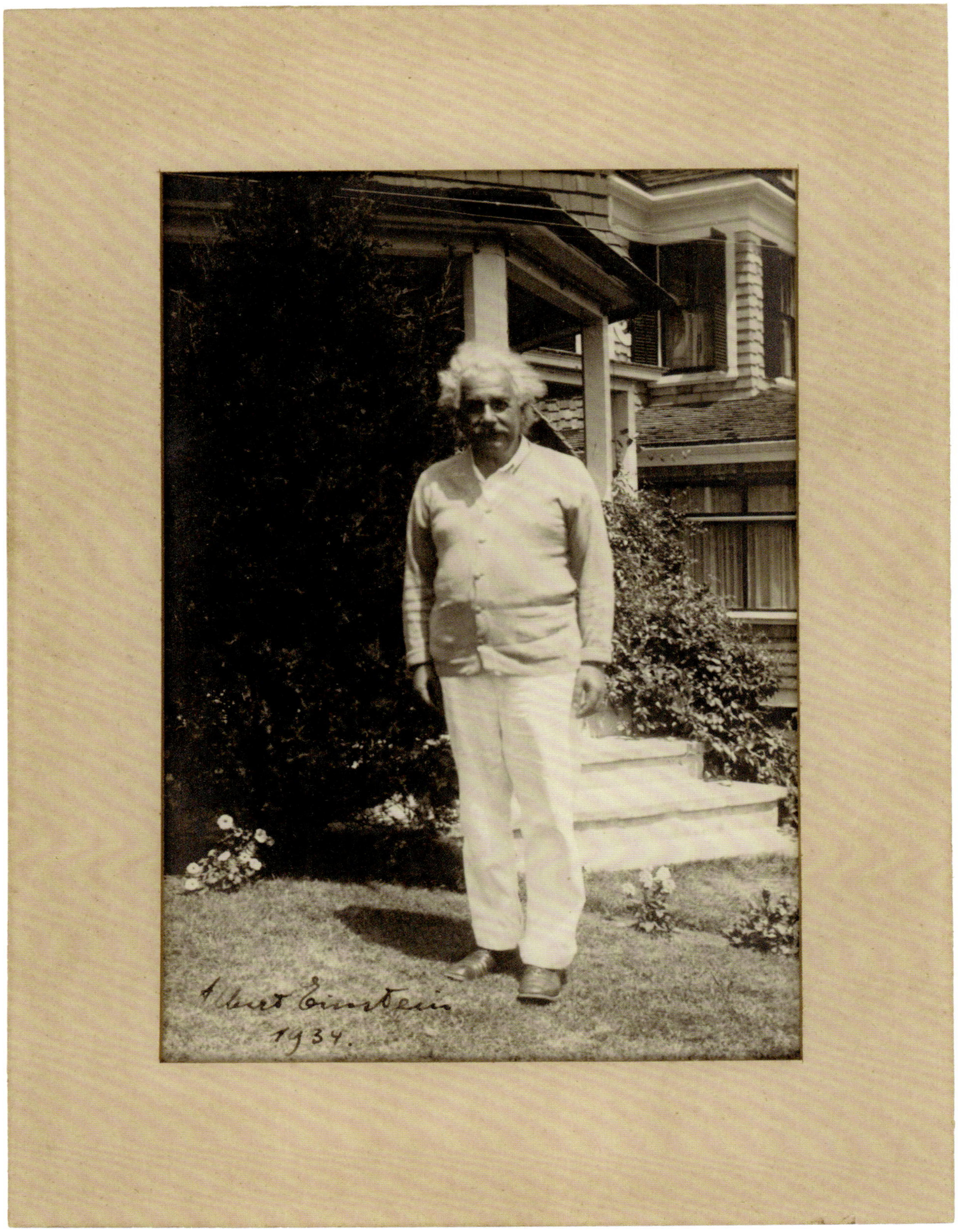
Albert Einstein
1934.

## *WATCH HILL*

After a tumultuous year, Einstein permanently settled in the United States. In 1933, he accepted a position, free of academic or teaching responsibilities, at the Institute for Advanced Study in Princeton. He would remain there for the rest of his life, walking to work from his home at 112 Mercer Street.

During summers, he and Elsa vacationed in the coastal town of Watch Hill, Rhode Island, where this photograph was taken.

"Einstein relished the simplicity of life in Watch Hill. He puttered around its lanes . . . [and] loved sailing his seventeen-foot wooden boat *Tinef*, which is Yiddish for a piece of junk. He usually went out on his own, aimlessly and often carelessly. 'Frequently he would go all day long, just drifting around,' remembered a member of a local yacht club who went to retrieve him on more than one occasion. 'He apparently was just out there meditating.'"

Einstein appears relaxed and sun-tanned in this candid photo, which he dated and signed in full.

UNKNOWN PHOTOGRAPHER,
*Rhode Island, 1934. Signed and dated by Einstein.*

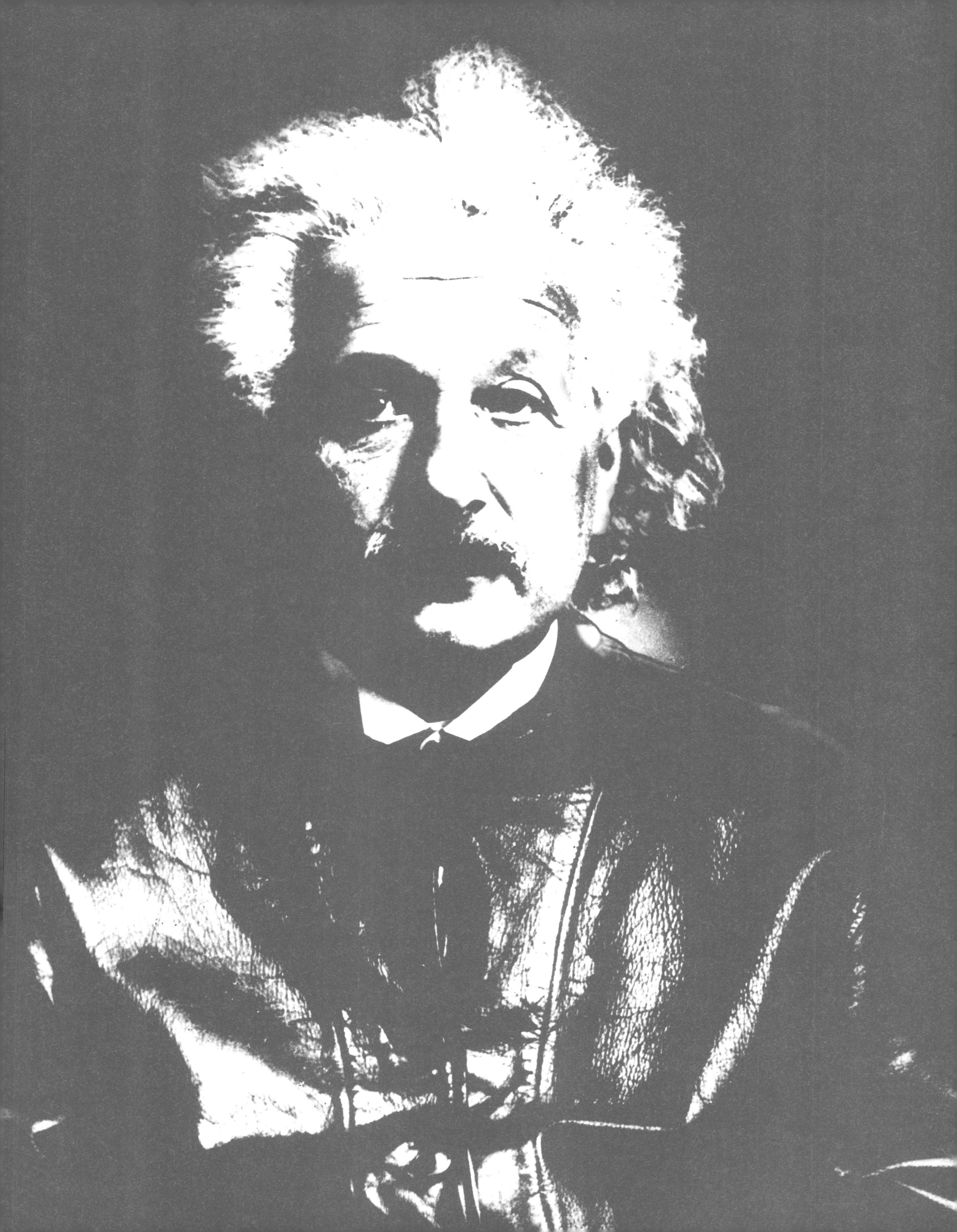

*As for the search for truth,*
*I know from my own*
*painful searching,*
*with its many blind alleys,*
*how hard it is to take a reliable step,*
*be it ever so small,*
*towards the understanding*
*of that which is truly significant."*

—ALBERT EINSTEIN, *1934*

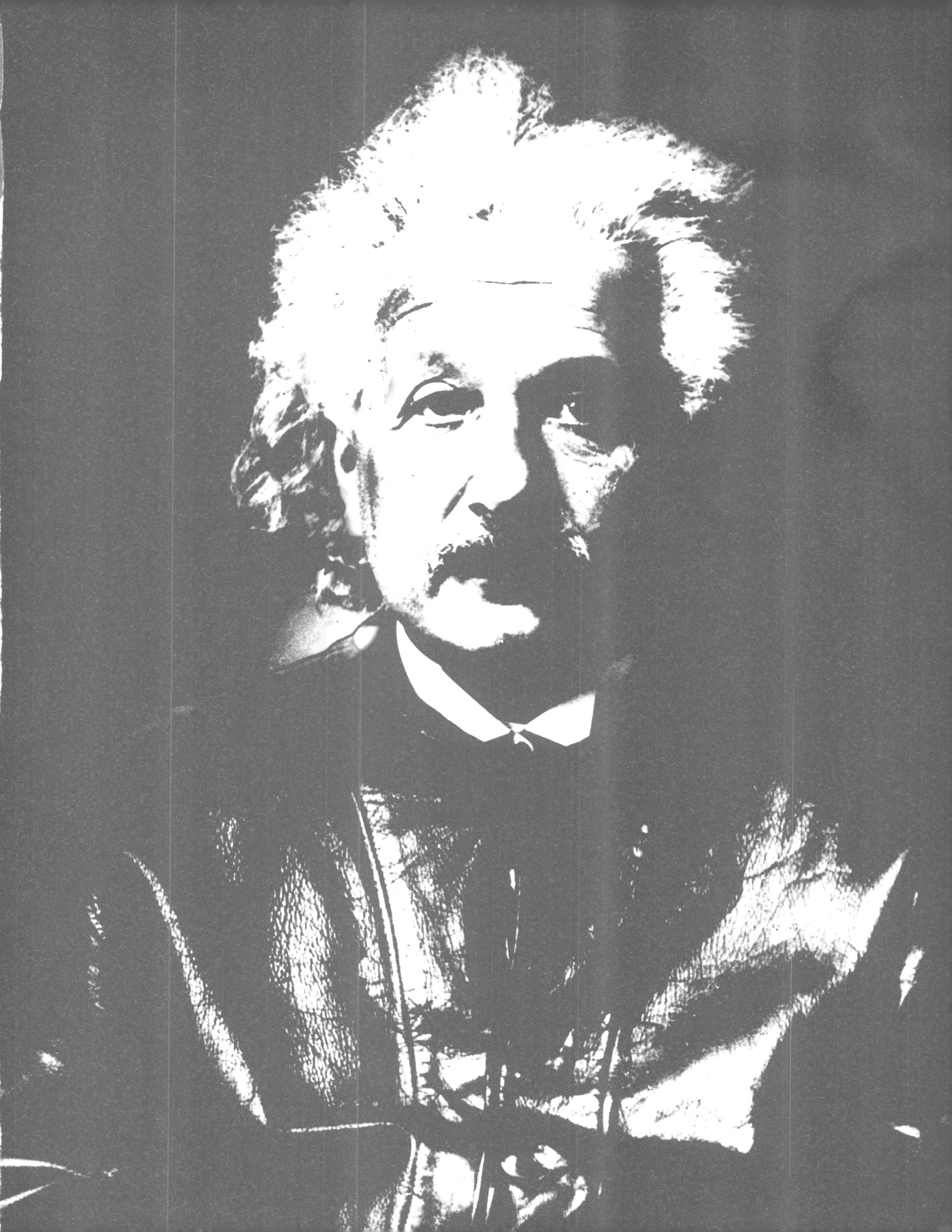

***WITH GOVERNOR A. HARRY MOORE***

In 1934, Governor A. Harry Moore of New Jersey visited Einstein at his Princeton home to invite him to a dinner, concert, and reception as an official welcome to the state. They appear to be enjoying their conversation in this photograph, which both men signed.

UNKNOWN PHOTOGRAPHER,
*Princeton, NJ, 1934. Signed by Einstein and New Jersey Governor A. Harry Moore.*

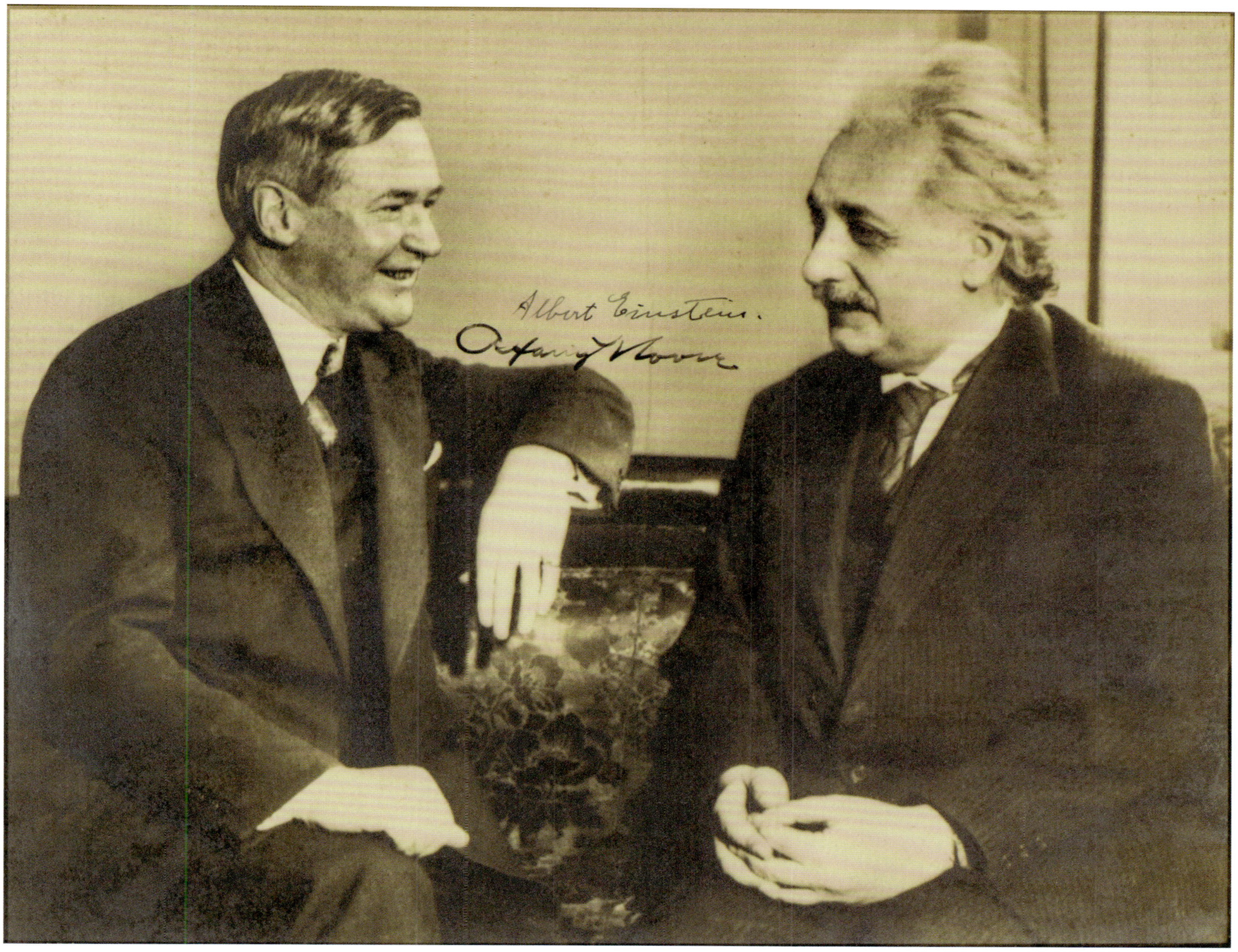
Albert Einstein.

## *THE QUEST FOR TRUTH*

This letter to Reinhold Pietsch of Utica, New York, dated February 13, 1934, contains one of Einstein's most often quoted remarks. Translated from German, it reads: "As for the search for truth, I know from my own painful searching, with its many blind alleys, how hard it is to take a reliable step, be it ever so small, towards the understanding of that which is truly significant."

Einstein said elsewhere that his supreme aim in life was the quest for truth. Here, he reminds us of the long and challenging path to gaining true understanding, even with the help of inspired thought experiments and intuition.

CORRESPONDENCE, *Einstein to Reinhold Pietsch of Utica, NY, 1934.*

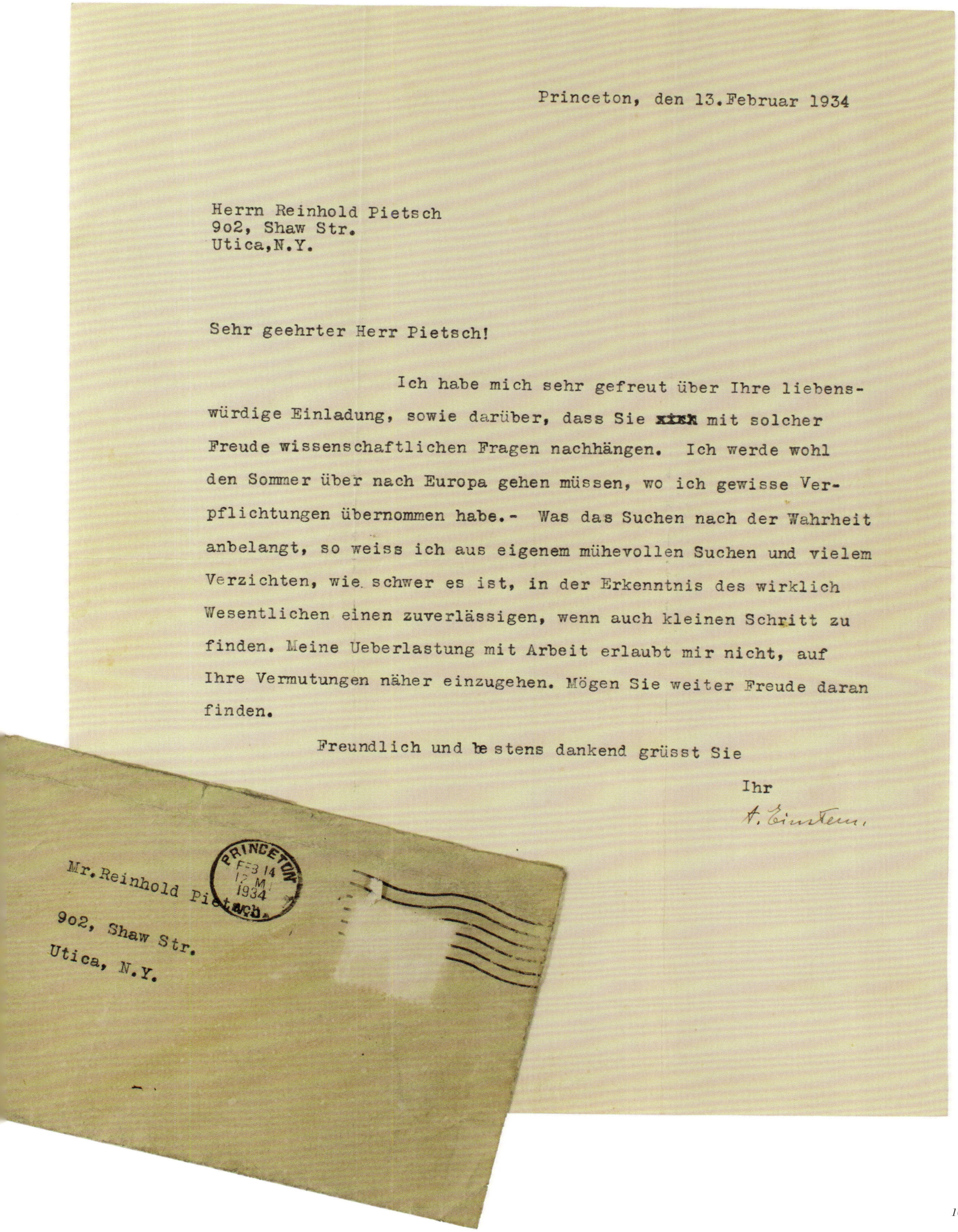

Princeton, den 13. Februar 1934

Herrn Reinhold Pietsch
9o2, Shaw Str.
Utica, N.Y.

Sehr geehrter Herr Pietsch!

Ich habe mich sehr gefreut über Ihre liebenswürdige Einladung, sowie darüber, dass Sie xixh mit solcher Freude wissenschaftlichen Fragen nachhängen. Ich werde wohl den Sommer über nach Europa gehen müssen, wo ich gewisse Verpflichtungen übernommen habe.- Was das Suchen nach der Wahrheit anbelangt, so weiss ich aus eigenem mühevollen Suchen und vielem Verzichten, wie schwer es ist, in der Erkenntnis des wirklich Wesentlichen einen zuverlässigen, wenn auch kleinen Schritt zu finden. Meine Ueberlastung mit Arbeit erlaubt mir nicht, auf Ihre Vermutungen näher einzugehen. Mögen Sie weiter Freude daran finden.

Freundlich und bestens dankend grüsst Sie

Ihr

A. Einstein.

## *SOCIETY CIRCUIT*

In the United States, Einstein was regarded not only as a scientist and philosopher but also a member of high society (no doubt to his amusement).

Here he is captured in formal attire by the New York–based Associated Press photographer Sophie Delar. Delar was known primarily for her "Society Circuit" photographs of singers and actors. Einstein inscribed and dated this photo on the photographer's matte.

PHOTOGRAPH BY SOPHIE DELAR,
*Princeton, NJ, 1935. Signed, dated, and inscribed by Einstein.*

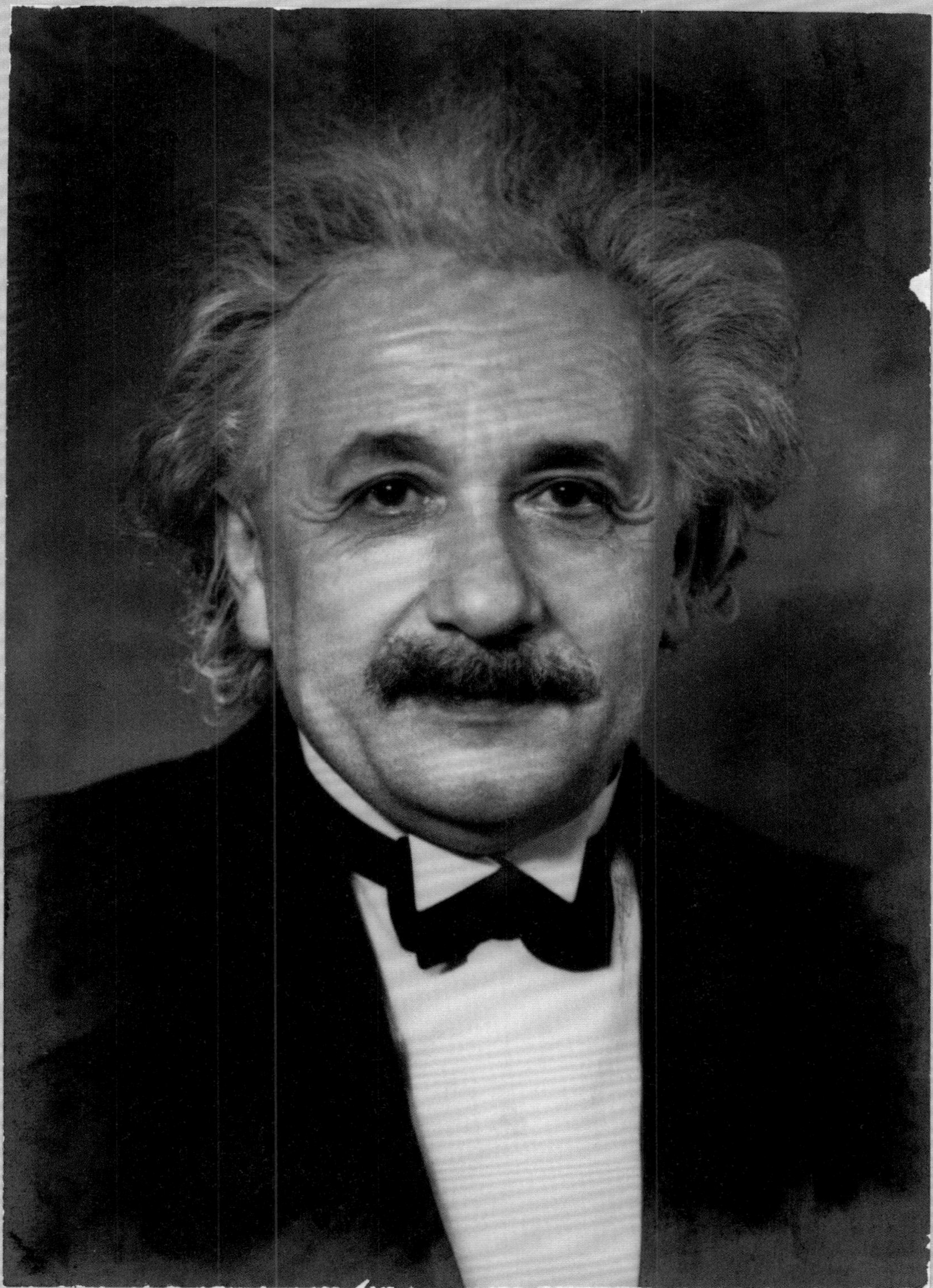

DELAR 30 ROCKEFELLER PLAZA

Zum Andenken an frohe musikalische Abende
Herrn und Frau Hermann Metzer

Albert Einstein 1935.

## *SPOOKY ACTION AT A DISTANCE*

In 1935, Einstein published a paper with coauthors Boris Podolsky and Nathan Rosen entitled, "Can Quantum-Mechanical Description of Physical Reality Be Considered Complete?" It became known as the EPR paper after the initials of the authors' last names. The paper described a thought experiment in which two quantum particles interact in a way that links their spatial coordinates and linear motion. (This new concept was later to be named "quantum entanglement.")

The authors described how determining the position (or momentum) of one particle would instantaneously identify the other particle's respective position (or momentum) without it being observed or measured. This contradicted the prevailing interpretation of quantum theory, based on the work of Niels Bohr and others, that maintained the atomic realm consists only of probabilities until measurements or observations are made. Einstein believed physical entities exist even when not observed, and their parameters can be determined precisely.

The argument introduced a paradox that could be resolved only if reality at the quantum level was not local or if information between the two conjoined particles traveled faster than the speed of light (an impossibility according to general relativity). Einstein believed that reality, even in the atomic world, was local. He disparagingly called the instantaneous interaction between the widely separated but linked particles "spooky action at a distance."

In 1964, nine years after Einstein's death, the physicist John Stewart Bell proposed a theorem disproving locality. He modified the EPR thought experiment so that it could be tested experimentally. Subsequent experimental data showed that quantum entanglement does exist and is non-local.

Although the EPR paper's conclusions were ultimately disproven, the concept of quantum particle entanglement introduced in it furthered the development of quantum theory. The EPR paper has become one of the most cited and discussed of all Einstein's works. Einstein continued challenging the indeterminism of quantum theory for the rest of his life, searching for an underlying, deterministic theory to unify quantum theory and general relativity.

*Pages 113–115:* **ALBERT EINSTEIN, BORIS PODOLSKY, AND NATHAN ROSEN,** *"Can Quantum-Mechanical Description of Physical Reality Be Considered Complete?" in* The Physical Review *47, no. 10, 1935.*

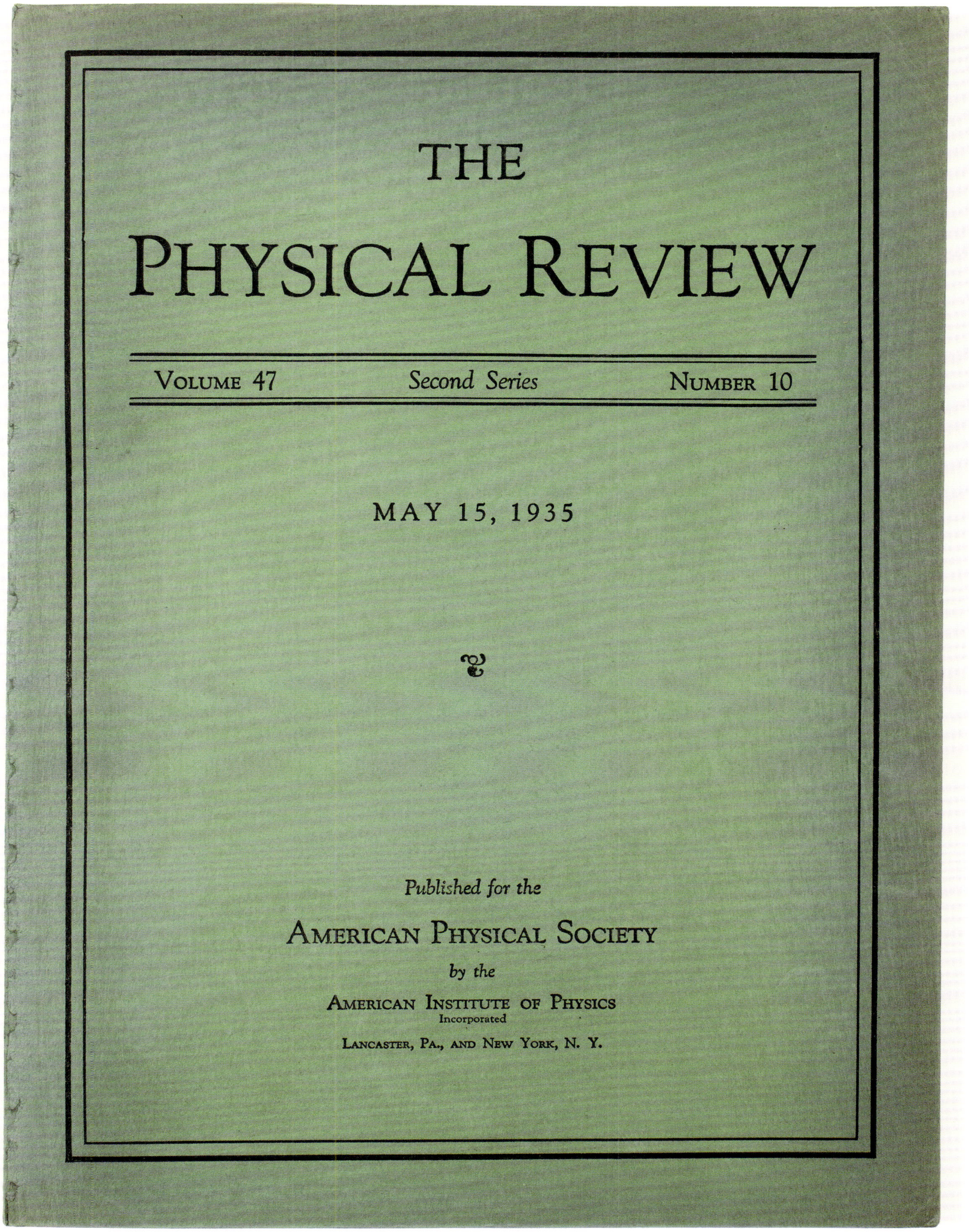

THE

PHYSICAL REVIEW

VOLUME 47 — *Second Series* — NUMBER 10

MAY 15, 1935

*Published for the*

AMERICAN PHYSICAL SOCIETY

*by the*

AMERICAN INSTITUTE OF PHYSICS

Incorporated

LANCASTER, PA., AND NEW YORK, N. Y.

approximations to the secular equations. They are

$$^4F_{9/2}:\quad K_1=0.9976,\qquad K_1^2=0.9952;$$
$$K_2=0.0693,\qquad K_2^2=0.0048.$$
$$^4F_{7/2}:\quad K_1=0.9966,\qquad K_1^2=0.9932;$$
$$K_2=-0.0313,\qquad K_2^2=0.00098;$$
$$K_3=-0.0762,\qquad K_3^2=0.0058.$$
$$^4F_{5/2}:\quad K_1=0.9951,\qquad K_1^2=0.9903;$$
$$K_2=0.0730,\qquad K_2^2=0.00533;$$
$$K_3=-0.0664,\qquad K_3^2=0.00441;\qquad K_4=0.$$
$$^4F_{3/2}:\quad K_1=0.9955,\qquad K_1^2=0.9911;$$
$$K_2=-0.0946,\qquad K_2^2=0.00894;$$
$$K_3=K_4=0.$$

The relativistic corrections for a $5d$ electron of La I, for which the effective nuclear charge is certainly less than 50, are very small and can be neglected. Thus for this application $a'$, $a''$, $a'''$ can be expressed in terms of $a_d$ by Eq. (22). When these substitutions are made and the values of the $K$'s listed above are inserted in Eqs. (23, 24, 25, 26) the following formulae for the interval factors of the $^4F$ states are obtained. They are equated on the right to Anderson's experimentally determined interval factors.

$$\begin{aligned} A(^4F_{9/2}) &= 0.1111a_s+0.6508a_d \\ &= 0.01571\ \text{cm}^{-1}, \\ A(^4F_{7/2}) &= 0.09514a_s+0.7864a_d \\ &= 0.01464\ \text{cm}^{-1}, \\ A(^4F_{5/2}) &= 0.03531a_s+1.0705a_d \\ &= 0.00900\ \text{cm}^{-1}, \\ A(^4F_{3/2}) &= -0.2000a_s+1.7988a_d \\ &= -0.01667\ \text{cm}^{-1}. \end{aligned} \tag{29}$$

There are four equations in $a_s$ and $a_d$, so we can solve for each and test the solution for consistency. Solving for $a_s$ and $a_d$ from the first and last equations of (29), which are the least sensitive to coupling changes, we obtain $a_s=0.1185\ \text{cm}^{-1}$, $a_d=0.00391\ \text{cm}^{-1}$. Substituting these values in the second and third equations of (29) we calculate $A(^4F_{7/2})=0.0143$, observed 0.0146; $A(^4F_{5/2})=0.0084$, observed 0.0090. Anderson gives the probable error of his most reliable interval factor as 5 percent. Thus we see that the preceding theory gives a consistent interpretation of the hyperfine separations of the $^4F$ states.

It is instructive to contrast the formulae for the interval factors of the $^4F$ states in strict $(LS)$ coupling with the formulae (29) which take into account the actual coupling conditions. The interval factors in strict $(LS)$ coupling are by Eqs. (21),

$$\begin{aligned} A(^4F_{9/2}) &= 0.1111a_s+0.6349a_d, \\ A(^4F_{7/2}) &= 0.07937a_s+0.7800a_d, \\ A(^4F_{5/2}) &= 0.00952a_s+1.0384a_d, \\ A(^4F_{3/2}) &= -0.2000a_s+1.7371a_d. \end{aligned} \tag{30}$$

Comparing (29) and (30), it is evident that $A(^4F_{9/2})$ and $A(^4F_{3/2})$ are not overly sensitive to departure from strict $(LS)$ coupling; $A(^4F_{7/2})$ is more sensitive, and $A(^4F_{5/2})$ is very sensitive. This comparison shows that one should be very careful in using $(LS)$ interval factor formulae for states dependent on coupling even though the multiplet structure indicates that the coupling is close to $(LS)$.

The nuclear magnetic moment of lanthanum can be calculated from the values of $a_s$ and $a_d$. The nuclear $g$ factor is computed from the interaction constant of an $s$ electron by the formula[8]

$$g(I)=\frac{3}{8}\,\frac{a_s}{R\alpha^2}\,\frac{n^3_{\text{eff.}}}{Z_iZ_0^2}\,\frac{1838}{K(\frac{1}{2},Z_i)}.$$

For our case $a_s=0.119\ \text{cm}^{-1}$, $n_{\text{eff.}}=1.60$, $Z_i=57$, $Z_0=1$, $R\alpha^2=5.82$, $K(\frac{1}{2},Z_i)=1.43$. The substitution of these values in the above equation gives $g(\text{I})=0.71$. A reliable value of $g(\text{I})$ cannot be readily obtained from $a_d$ since first, $a_d$ is very small, and second, it is difficult to estimate the value of $Z_i$ that should be used for a $5d$ electron in the $5d^26s$ configuration. However, the substitution of $g(I)=0.71$, $a_d=0.0039\ \text{cm}^{-1}$, $\Delta\nu=1000\ \text{cm}^{-1}$ in the formula for a non-$s$ electron[8] gives, on solving for $Z_i$, $Z_i=40$. This appears to be a reasonable value, indicating that $g(\text{I})=0.71$ is consistent with $a_d=0.0039$. The nuclear spin

[8] S. Goudsmit, Phys. Rev. **43**, 636 (1933); E. Fermi and E. Segrè, Zeits. f. Physik **82**, 729 (1933).

of lanthanum is 7/2, hence the nuclear magnetic moment as determined by this analysis is 2.5 nuclear magnetons. This is in fair agreement with the value 2.8 nuclear magnetons determined from La III hyperfine structures by the writer and N. S. Grace.[9]

[9] M. F. Crawford and N. S. Grace, Phys. Rev. **47**, 536 (1935).

This investigation was carried out under the supervision of Professor G. Breit, and I wish to thank him for the invaluable advice and assistance so freely given. I also take this opportunity to acknowledge the award of a Fellowship by the Royal Society of Canada, and to thank the University of Wisconsin and the Department of Physics for the privilege of working here.

---

MAY 15, 1935 PHYSICAL REVIEW VOLUME 47

# Can Quantum-Mechanical Description of Physical Reality Be Considered Complete?

A. Einstein, B. Podolsky and N. Rosen, *Institute for Advanced Study, Princeton, New Jersey*

(Received March 25, 1935)

In a complete theory there is an element corresponding to each element of reality. A sufficient condition for the reality of a physical quantity is the possibility of predicting it with certainty, without disturbing the system. In quantum mechanics in the case of two physical quantities described by non-commuting operators, the knowledge of one precludes the knowledge of the other. Then either (1) the description of reality given by the wave function in quantum mechanics is not complete or (2) these two quantities cannot have simultaneous reality. Consideration of the problem of making predictions concerning a system on the basis of measurements made on another system that had previously interacted with it leads to the result that if (1) is false then (2) is also false. One is thus led to conclude that the description of reality as given by a wave function is not complete.

## 1.

Any serious consideration of a physical theory must take into account the distinction between the objective reality, which is independent of any theory, and the physical concepts with which the theory operates. These concepts are intended to correspond with the objective reality, and by means of these concepts we picture this reality to ourselves.

In attempting to judge the success of a physical theory, we may ask ourselves two questions: (1) "Is the theory correct?" and (2) "Is the description given by the theory complete?" It is only in the case in which positive answers may be given to both of these questions, that the concepts of the theory may be said to be satisfactory. The correctness of the theory is judged by the degree of agreement between the conclusions of the theory and human experience. This experience, which alone enables us to make inferences about reality, in physics takes the form of experiment and measurement. It is the second question that we wish to consider here, as applied to quantum mechanics.

Whatever the meaning assigned to the term *complete*, the following requirement for a complete theory seems to be a necessary one: *every element of the physical reality must have a counterpart in the physical theory.* We shall call this the condition of completeness. The second question is thus easily answered, as soon as we are able to decide what are the elements of the physical reality.

The elements of the physical reality cannot be determined by *a priori* philosophical considerations, but must be found by an appeal to results of experiments and measurements. A comprehensive definition of reality is, however, unnecessary for our purpose. We shall be satisfied with the following criterion, which we regard as reasonable. *If, without in any way disturbing a system, we can predict with certainty (i.e., with probability equal to unity) the value of a physical quantity, then there exists an element of physical reality corresponding to this physical quantity.* It seems to us that this criterion, while far from exhausting all possible ways of recognizing a physical reality, at least provides us with one

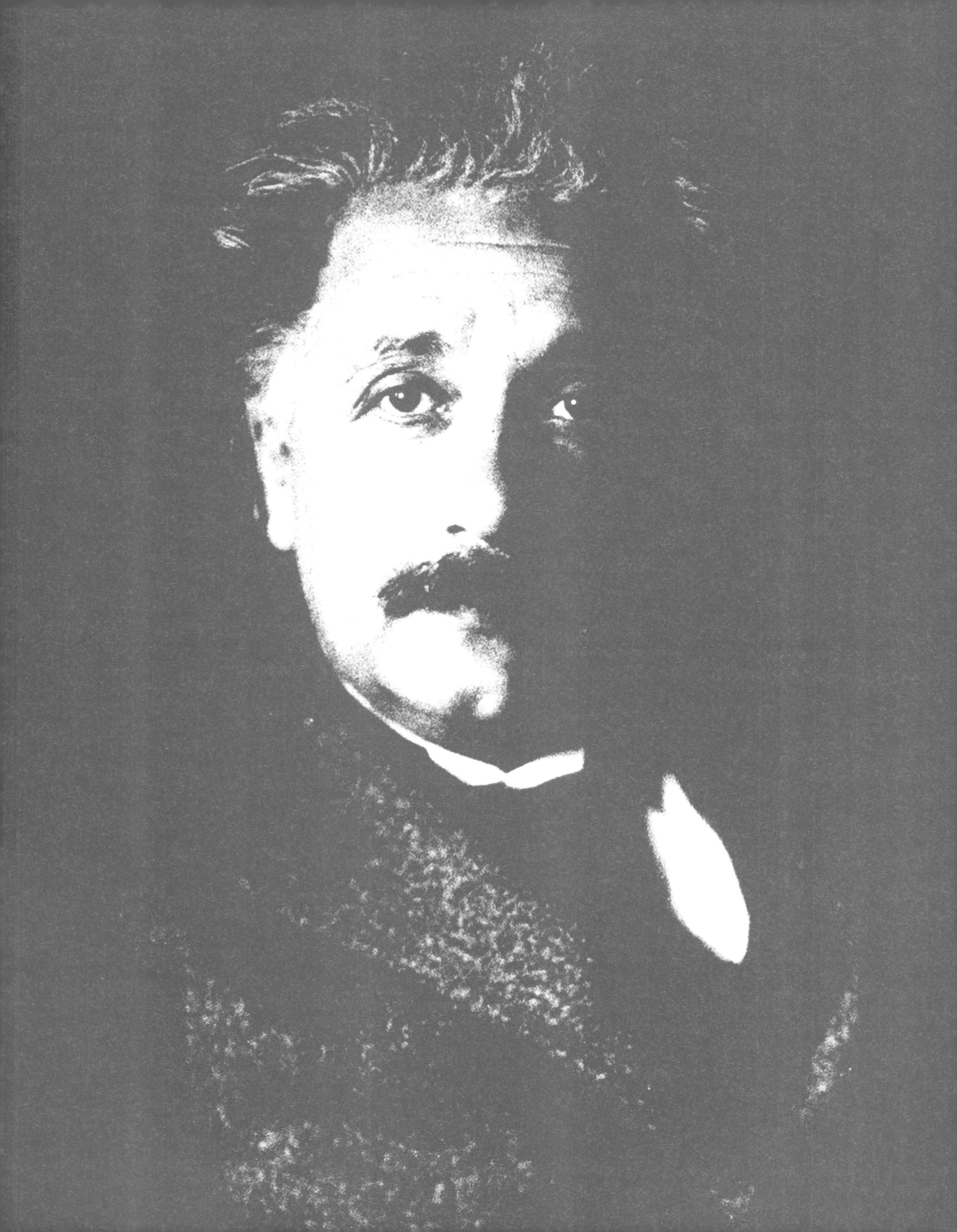

*The following requirement for a complete theory seems to be a necessary one: every element of the physical reality must have a counterpart in the physical theory."*

—ALBERT EINSTEIN, *1935*

er den Tropfen bildenden

$= \lg \mu_0 + \alpha' \Delta$ . . . . (1)

ringe Modifikationen hand

$\mu_0 = \lg \frac{\mu}{\mu_0} = \lg \frac{\mu_0 + (\mu - }{\mu_0}$

t ergibt dies mit genüger

$\alpha' \mu_0 \Delta$, oder $\boxed{\mu = \mu_0 + A}$

ndere Konstante $(\alpha' \mu_0)$

sind wir ferner darüber

## *THE PHENOMENA OF CAPILLARITY*

In his *Autobiographical Notes*, published in 1949, Einstein reflected on his youthful scientific passions and singled out his early fascination with thermodynamics: "A theory is the more impressive the greater the simplicity of its premises is, the more different kinds of things it relates, and the more extended is its area of applicability. Therefore the deep impression which classical thermodynamics made upon me. It is the only physical theory of universal content concerning which I am convinced that, within the framework of applicability of its basic concepts, it will never be overthrown."

Einstein's first published paper, "Conclusions Drawn from the Phenomena of Capillarity," was on the subject of thermodynamics. It appeared in *Annalen der Physik* in 1901, when the author was just twenty-one years old.

In this 1937 letter to Hans M. Cassel, a lecturer at the Technical University of Berlin, Einstein returns to the subject of thermodynamics. The letter is a detailed description, accompanied by handwritten equations and formulas, of the pressure and radius requirements to achieve stable equilibrium for droplets subjected to capillary action.

Thermodynamics played a significant role in Einstein's thinking throughout his career. Its principles are embedded in his writings on molecular motion, Bose-Einstein statistics, and quantum mechanics. It is no exaggeration to say that "all of Einstein's boldly original attacks on what he saw as the critical problems of early-20th-century physics are intimately related to his understanding of thermodynamics."

*Pages 118, 120–121:* CORRESPONDENCE,
*Einstein to Hans M. Cassel, 1937.*

Princeton N.J., den 27.Januar 1937
112 Mercer Str.

Herrn Hans Cassel
195 Park Boulevd.
Palo Alto, Cal.

Sehr geehrter Herr Cassel:

Nun habe ich doch Hoffnung, dass wir die Sache ins Klare bringen. Sie haben mir nämlich wirklich auf meinen Brief geantwortet. Ich werde Ihnen nun genau sagen, inwiefern ich Ihrer Ueberlegung nicht folgen kann.

1). Einig sind wir (in Ihrer Fassung) darüber: Lastet auf der einen Tropfen umgebenden semipermeablen Membran ein äusserer Druck $\Delta$, so steigt der Dampfdruck der den Tropfen bildenden Flüssigkeit von $p_0$ auf p gemäss der Formel

$$\lg p = \lg p_0 + a'\Delta \quad \ldots (1)$$

Da es sich um geringe Modifikationen handelt, kann man setzen

$$\lg p - \lg p_0 = \lg \frac{p}{p_0} = \lg \frac{p_0 + (p - p_0)}{p_0} = \lg \left(1 + \frac{p - p_0}{p_0}\right) = \frac{p - p_0}{p_0}$$

In (1) eingesetzt ergibt dies mit genügender Näherung

$$p = p_0 + a' p_0 \Delta, \text{ oder } \boxed{p = p_0 + A\Delta} \quad \ldots (1a)$$

wobei $A$ eine andere Konstante $(a' p_0)$ bedeutet.

2.) Einig sind wir ferner darüber, dass für $\Delta$ der Kapillardruck $\frac{2\sigma}{r}$ einzusetzen ist, sodass wir also haben

$$p = p_0 + 2A\frac{\sigma}{r} \quad \ldots\ldots (1b)$$

Punkt, in welchem ich
die Stabilität eines
abil ist, wenn

grösserem Radius $(r+\Delta r)$
össer als $p$ ist, und für
$r$ der Dampfdruck kleiner ist
ius $r$ durch Verdampfung, bezw.

Bedingung

Sie verlangen, dass $\frac{dp}{dr}$
äre ja das Gleichgewicht
g für das stabile Gleichgewicht.
tpunkte der Uneinigkeit. Damit
und Gleichung (1 b) möglich sei, muss

ichen Gründen das Gleichheitszeichen).
, dass die Erfüllung dieser Un-
voraussetze. Sie behaupten, dass diese
er kapillar-aktiven Substanz tat-

e ich.
hierfür angeführt haben, sind mir
irklich behaupten, dass die Flächen-
von deren Partialdruck auch noch

radius merklich abhänge? Wenn nicht, so folgt aus der
Gleichung

$$-d\sigma = \Gamma R T d(\lg p'),$$

llein abhängt. Man braucht nur in die Gibbs'sche
Funktion von $p'$ eingesetzt zu denken und zu in-
aber $\sigma$ von $p$ unabhängig ist, so ist die Bedingung
lität niemals erfüllt.
t-Druck im Innern des Tropfens erfüllt nun aller-

$$P = dp + dp' + d\left(\frac{2\sigma}{r}\right)$$

edingung der Tropfen-Stabilität nicht das Geringste
ann mich Ihren diesbezüglichen Erwägungen nicht

ch grüsst Sie

Ihr

A. Einstein.

## *RELATIVE TIDE AND SAND BARS*

Long Island was another place where Einstein enjoyed summers in his sailboat deep in thought, often to the detriment of his safety. After one mishap in the Long Island Sound, the *New York Times* ran a front-page article with the headline, "Relative Tide and Sand Bars Trap Einstein."

This dramatic photo of a handsome, deeply tanned, fifty-eight-year-old Einstein was taken in Huntington, Long Island, by Lotte Jacobi and signed by the photographer. Jacobi was known for her high-contrast, intimate portraits. Her approach was to "get people to talk, to relax, to be themselves . . . to bring out their personalities."

Jacobi had previously photographed Einstein in 1927 at her father's Berlin studio. Like Einstein, she fled Germany in the 1930s. The two became lifelong friends. Einstein called Lotte Jacobi his favorite photographer.

PHOTOGRAPH BY LOTTE JACOBI,
*Long Island, NY, 1937. Signed by the photographer.*

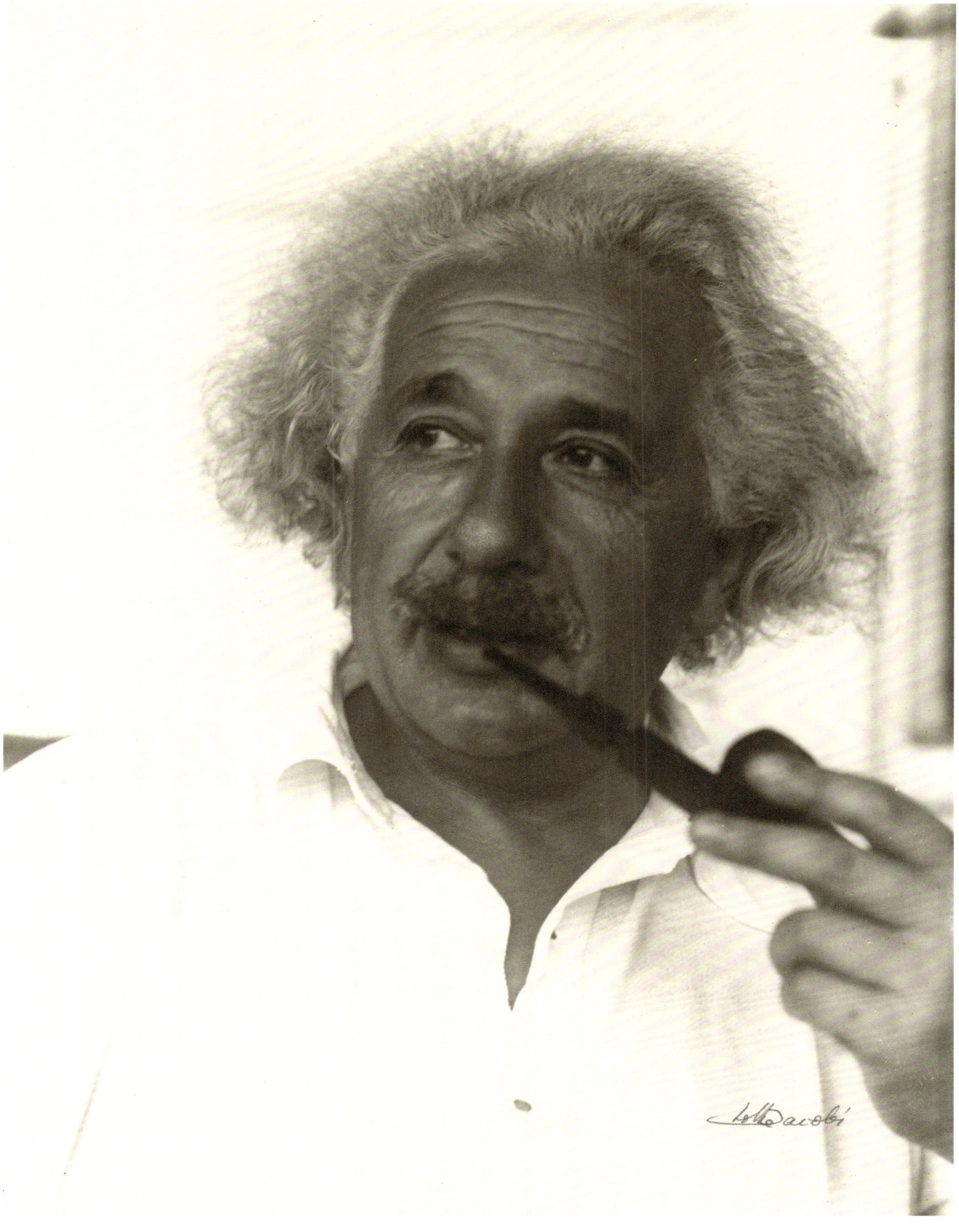

## *IN HIS STUDY*

Einstein agreed to sit for a photo story by *Life* magazine in 1938, provided that the photographer was Lotte Jacobi.

Jacobi photographed many of the twentieth century's most famous men and women. She was descended from a long line of photographers. Her great-grandfather learned the craft from the inventor of photography himself, Louis Daguerre. Being Jewish, Jacobi abandoned her successful career in Germany in 1935 after rejecting the Nazis' offer of "honorary Aryan status." This was undoubtedly one of the reasons Einstein so admired her.

Jacobi disliked studio techniques that produced stiff portraiture. Instead, she brought the studio to her subjects, talked with them, and waited patiently until they were entirely at ease before taking photos. Here she exquisitely captured Einstein at his Princeton home in a private moment of deep thought, working on scientific equations, his hair untamed, and dressed casually in his favorite leather bomber jacket.

*Life* magazine rejected the photograph on the grounds that it defied the prevailing aesthetics of photojournalism and did not treat Einstein "with enough respect." But in focusing on the very qualities that defined the inner nature of her subject—capturing his meditative absorption in his science and his indifference to social conventions of dress and grooming—Jacobi created one of the most famous and beloved images of this extraordinary man.

The photo was first publicly displayed at the Museum of Modern Art's 1942 exhibition *20th Century Portraits.* It has since become an important image, defining our conception of the great scientist.

The print shown here is signed and dated by Einstein on the original presentation matte and is signed by Jacobi.

PHOTOGRAPH BY LOTTE JACOBI,
*Princeton, NJ, 1938. Signed and dated by Einstein in 1949 and signed by the photographer.*

*Long hair minimizes*
*the need for barbers;*
*socks can be done without;*
*one leather jacket solves*
*the coat problem for*
*many years; suspenders*
*are superfluous."*

—LEOPOLD INFELD ON EINSTEIN, *1929*

Long hair minimizes
the need for barbers;
socks can be done without;
one leather jacket solves
the coat problem for
many years; suspenders
are superfluous.”

—LEOPOLD INFELD ON EINSTEIN, 1929

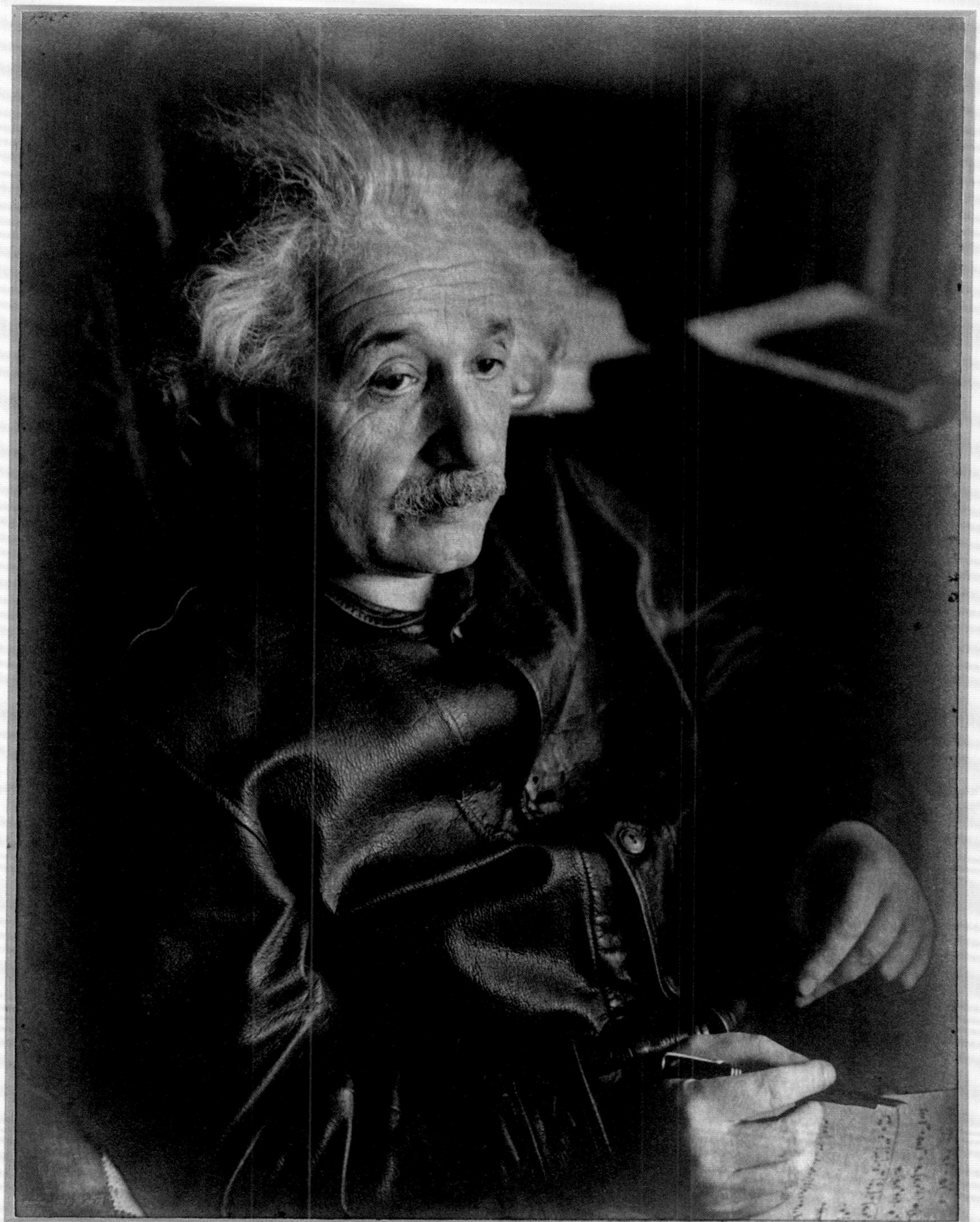

A. Einstein 19.IV.49.

Jacobi

## *LEATHER BOMBER JACKET*

At Princeton, Einstein collaborated with many other Jewish physicists who escaped the Nazis. One of them was Leopold Infeld, who noted of Einstein, "We are slaves of bathrooms, Frigidaires, cars, radios, and millions of other things. Einstein tried to reduce them to the absolute minimum. Long hair minimizes the need for barbers; socks can be done without; one leather jacket solves the coat problem for many years."

The Levi leather jacket, which Jacobi's photographs helped make famous, was one of Einstein's most beloved possessions. He bought it soon after settling in the United States and said it made him feel more like a true American.

This image of Einstein relaxing in his study in his leather jacket was taken during the celebrated 1938 photo session with Jacobi, commissioned by *Life* magazine. Einstein signed on the image over his left hand.

PHOTOGRAPH BY LOTTE JACOBI,
*Princeton, NJ, 1938. Signed by Einstein.*

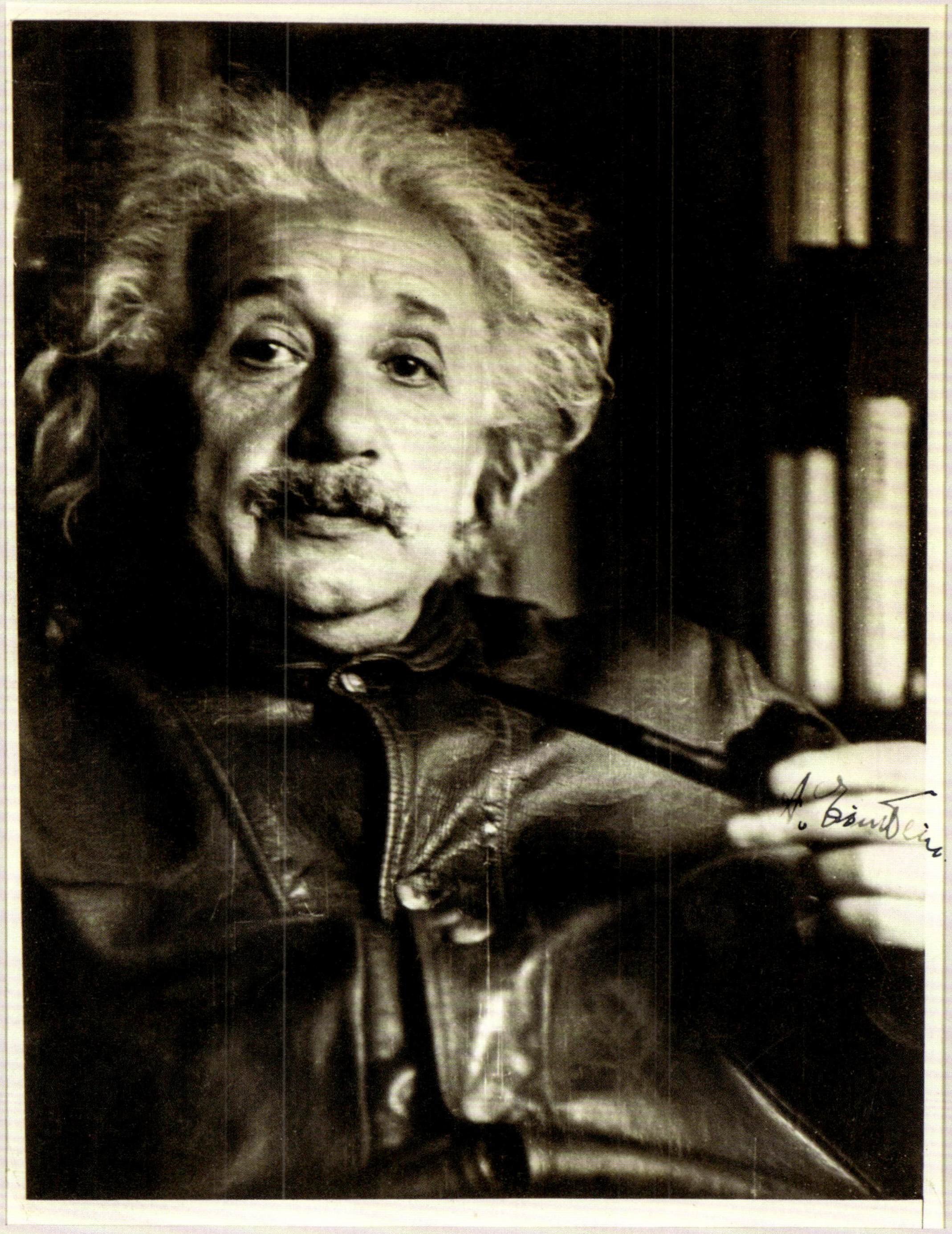
A. Einstein

### *LIFE MAGAZINE*

This "respectful" image of Einstein is the one *Life* magazine's editors chose for its April 11, 1938, issue, printed in color to accompany the article "Albert Einstein Simplifies Relativity."

It is among the most dramatic images of Einstein ever produced. The photographer captured him from the side, with his head turned toward the camera and his eyes slightly downcast. Smoke from his pipe creates a hazy backdrop, emphasizing the crisp lines of his face and hair. These elements give Einstein a majestic—almost ethereal—look. We have not determined which of the *Life* magazine photographers took this photograph.

Einstein inscribed in ink above the bowl of his pipe: "Herr Reves / A. Einstein / 1938."

"Herr Reves" was Hungarian-born Emery Reves, an influential figure in Einstein's life. A writer, economist, literary agent for Winston Churchill, and political operative, Reves shared Einstein's passionate dedication to pacifism and anti-Nazism. He published an influential book in 1945, *The Anatomy of Peace*, arguing for the cause of world federalism. Einstein wholeheartedly endorsed the book, writing to the American physicist J. Robert Oppenheimer on September 29, 1945: "Mr. Emery Reves, whom I have known for many years and with whom I have often discussed urgent political problems, has sent me a copy of the statement which you and your colleagues issued for the enlightenment of the public and the government. While I was very much pleased by the candid language and the sincerity of the statement, I was, at the same time, somewhat bewildered by the political recommendations, which I consider inadequate. . . . At the present high level of industrialization and economic interdependence, it is unthinkable that we can achieve peace without a genuine supranational organization to govern international relations. If war is to be avoided, anything less than such an overall solution strikes me as illusory.

"A few weeks ago, Emery Reves published a short book entitled *The Anatomy of Peace* which, in my opinion, explains the problem as clearly and pertinently as anyone ever has. I have learnt that several men who play an active role in public life are taking steps to make the book known to every American. I urge you and your colleagues to read it and discuss its conclusions. Although it was written before the explosion of the atomic bomb, it contains a solution which is directly applicable to the problem created by this new weapon. I shall be glad to send you a number of copies for distribution or to mail copies directly to you and your colleagues if you send me their addresses. I am convinced that the political part of the statement which you and your colleagues issued would have been formulated differently, had the facts and discussions presented in this book been made known to who drafted the statement."

UNKNOWN PHOTOGRAPHER,
*United States, 1938. Signed, dated, and inscribed by Einstein: "Herr Reves / A. Einstein / 1938."*

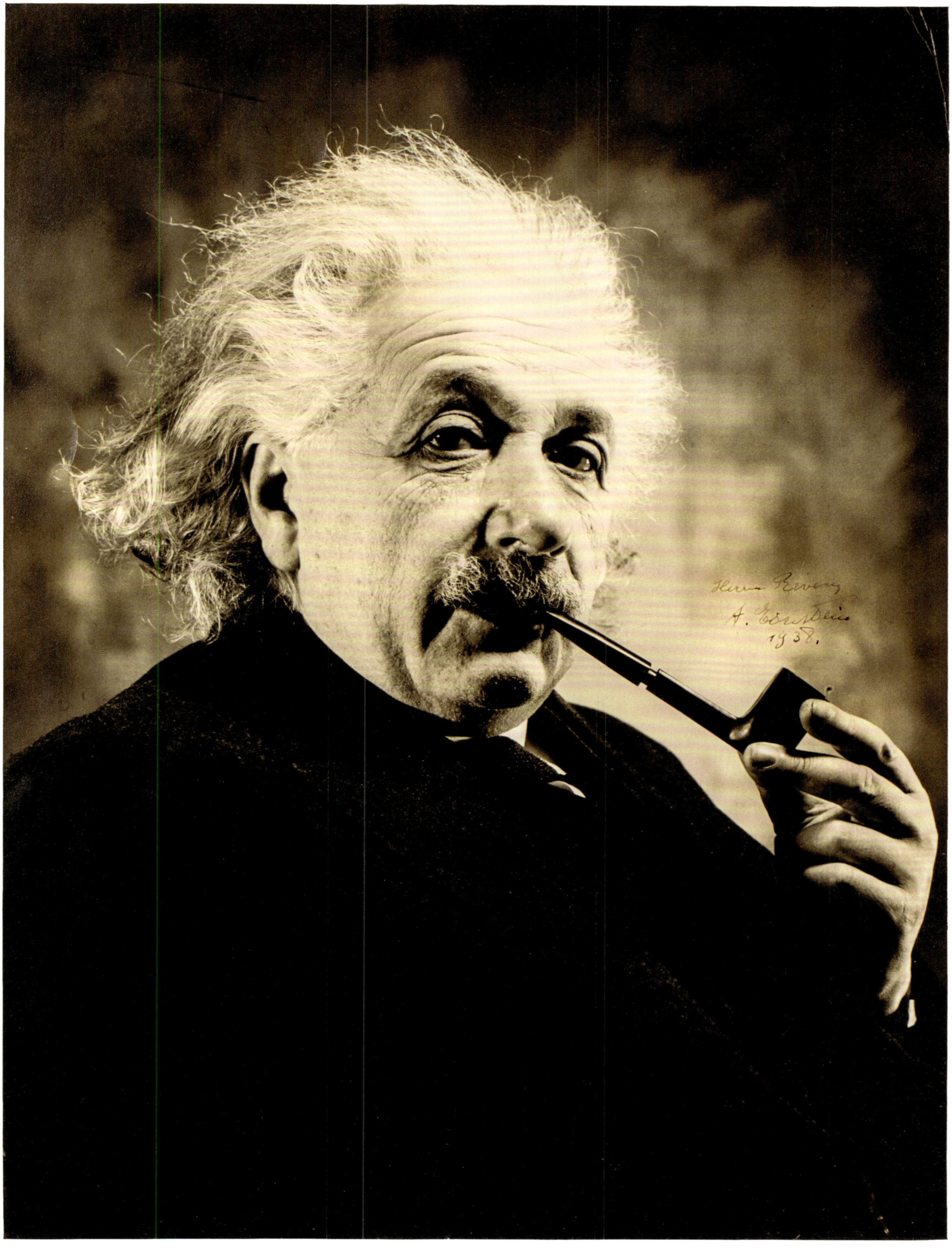
A. Einstein
1938.

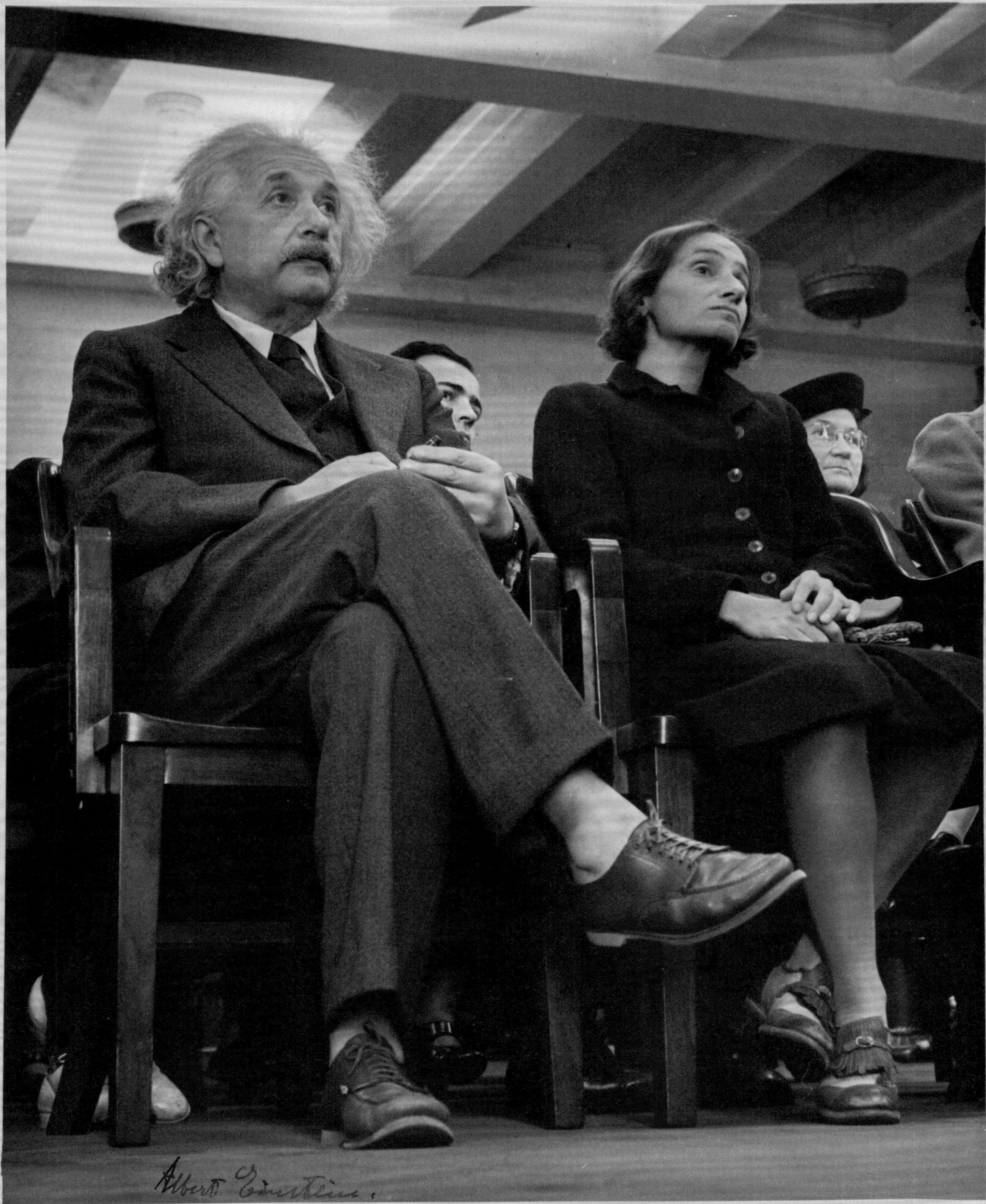
Albert Einstein.

## *BECOMING A UNITED STATES CITIZEN*

Einstein "took his citizenship test on June 22, 1940, in front of a federal judge in Trenton. . . . [He] passed his test and he was sworn in—along with his stepdaughter Margot, his assistant Helen Dukas, and eighty-six other new citizens—on October 1. Afterward, he praised America to the reporters covering his naturalization. The nation, he said, would prove that democracy is not just a form of government but 'a way of life tied to a great tradition, the tradition of moral strength.' Asked if he would renounce other loyalties, he joyously declared that he 'would even renounce my cherished sailboat' if that were necessary. . . .

"When he first arrived in Princeton, Einstein had been impressed that America was, or could be, a land free of the rigid class hierarchies and servility in Europe. But what grew to impress him more—and what made him fundamentally such a good American but also a controversial one—was the country's tolerance of free thought, free speech, and nonconformist beliefs. That had been a touchstone of his science, and now it was a touchstone of his citizenship."

A photographer for a Trenton, New Jersey, newspaper took this photograph of Einstein at his naturalization ceremony. Einstein's stepdaughter, Margot, is seated in the chair next to him. The low camera position and upward angle amusingly emphasize that Einstein was sockless even on such a formal occasion.

A day or two later, the photographer visited Einstein in Princeton and asked him to sign the print. The new United States citizen obliged, writing his full name in ink in the lower-left corner.

UNKNOWN PHOTOGRAPHER, *Trenton, NJ, 1940. Einstein at his American naturalization ceremony. Signed by Einstein.*

*“I have no special talent.*
*I am only passionately curious.”*

—ALBERT EINSTEIN, *1952*

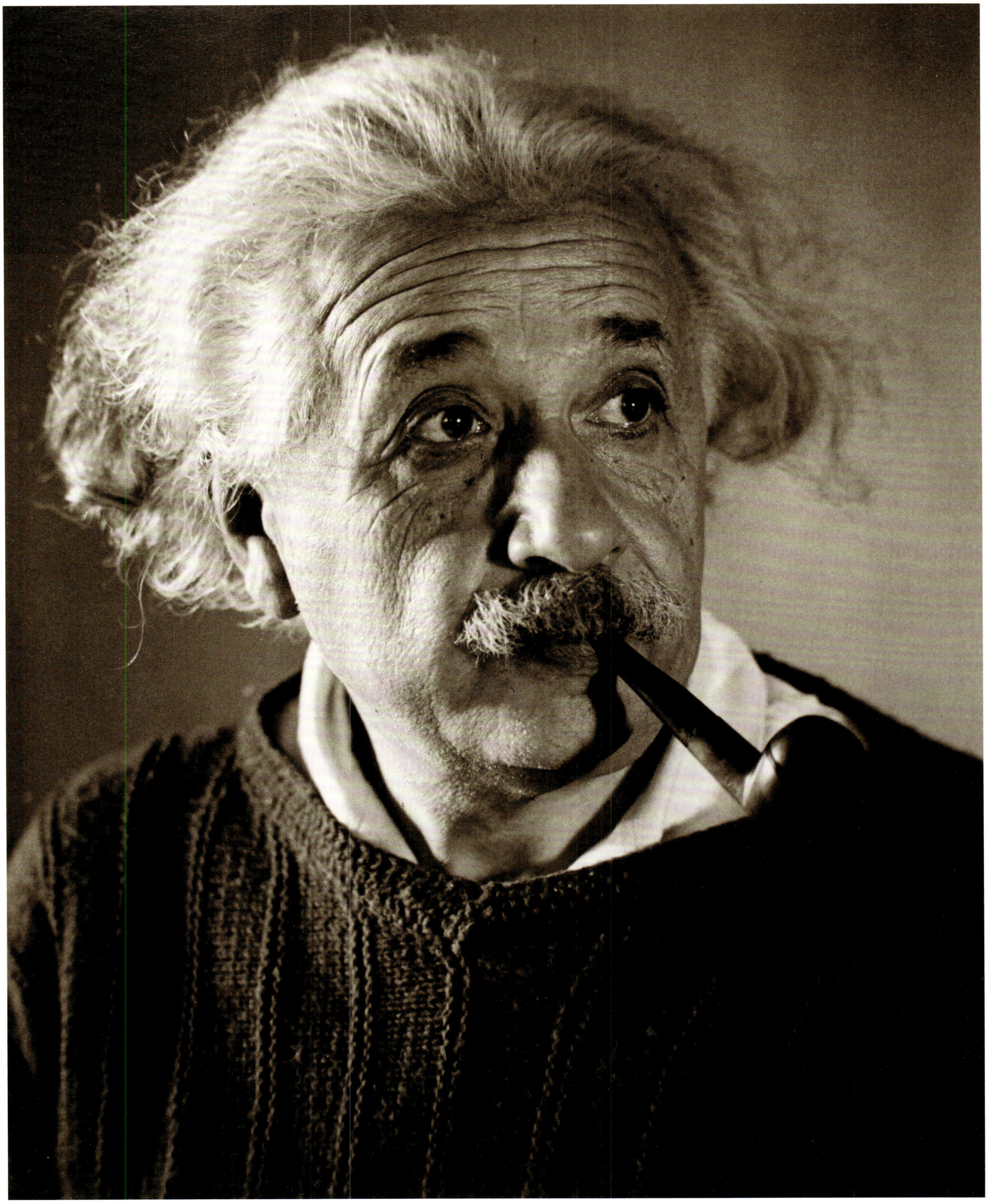

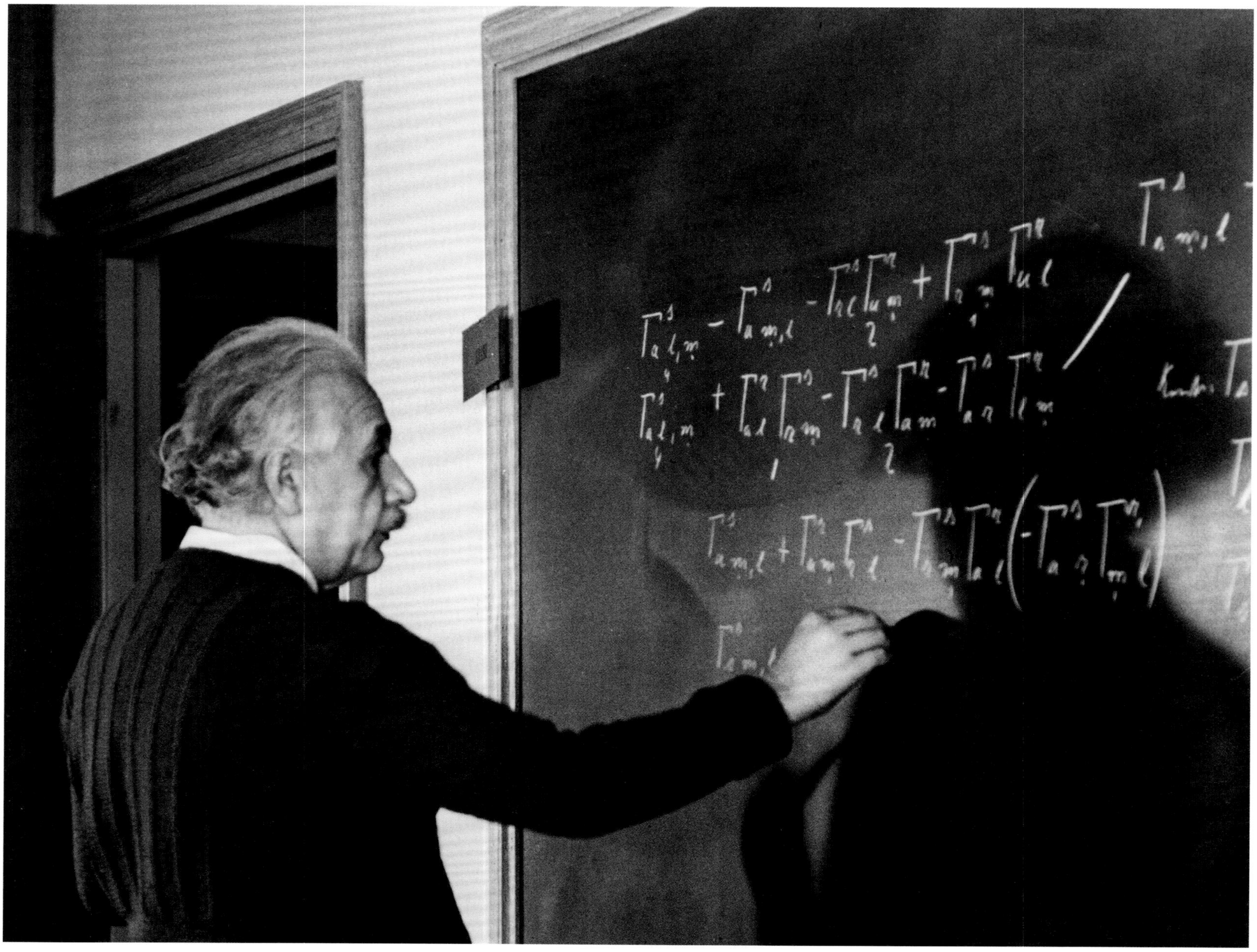

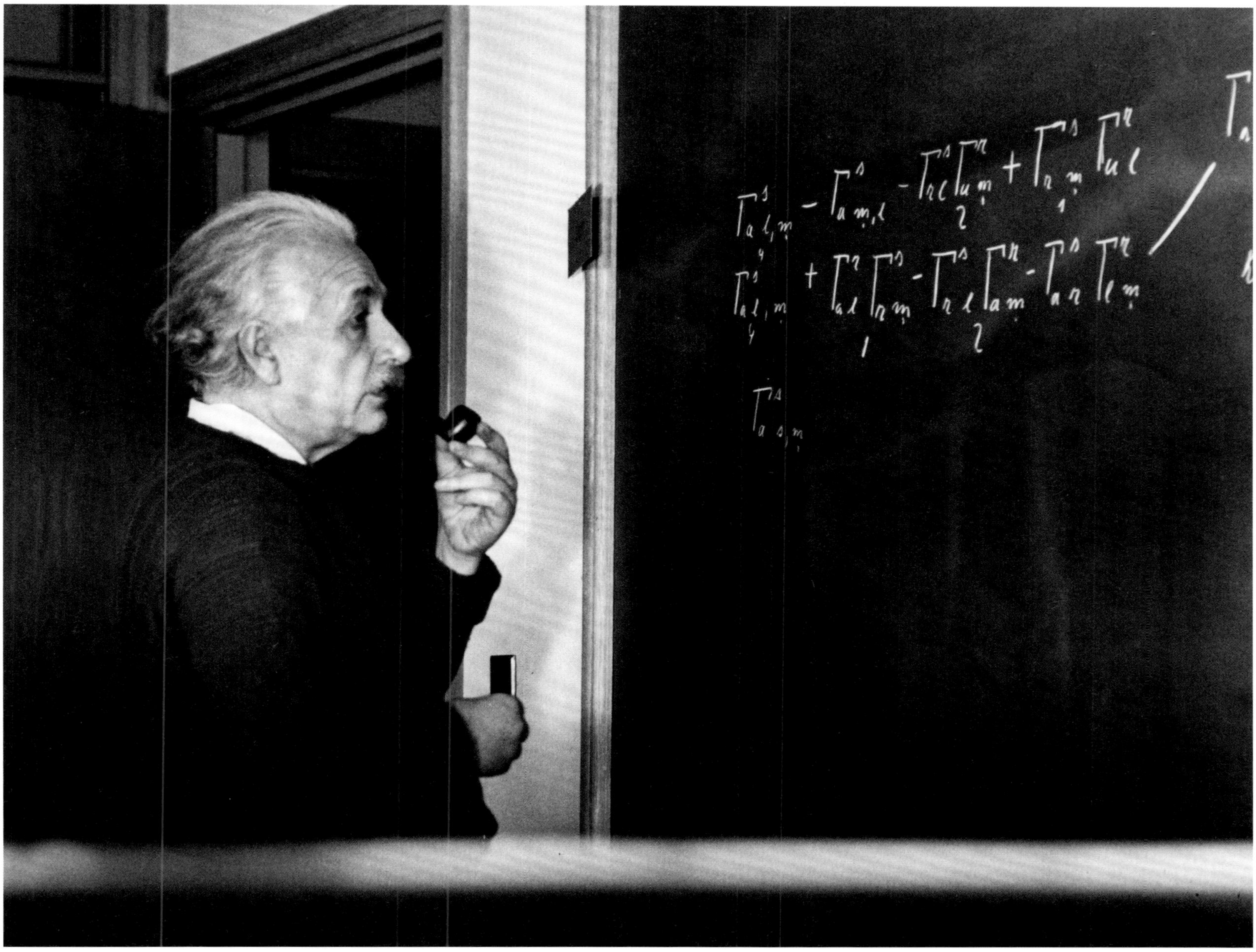

## *EINSTEIN AT WORK: A SUITE OF SEVEN PHOTOGRAPHS*

An uninvited visitor arrived at Einstein's home one day with greetings from mutual friends in Berlin. The visitor was a Russian American photographer, Roman Vishniac, famous for his photographs of Eastern European Jewish communities preceding the Holocaust. Fleeing war-torn Europe, Vishniac had arrived in the United States in 1940. He hoped Einstein would pose for a portrait, but Einstein had little interest in doing so.

Vishniac recalled his visit with Einstein: "It was a singular experience. An idea had suddenly come to him, and the room was filled with the movement of the great man's thought. I waited several minutes, and then when I saw that he did not intend to say anything more to me and that he was off in a world of his own, I started taking pictures."

Vishniac assembled seven oversized photos in a book-like case stamped "Einstein at Work" and "Roman Vishniac." He signed and dated each print "Princeton 1942" below the image and numbered the backs of the prints.

Vishniac added a handwritten letter to the original owner, dated July 1980, describing the circumstances of the photoshoot: "The originality of Portfolio 'Einstein' consists of its special character. It is made not to get images, but the feeling that you are present during the creativity by the Great Man. All pictures are made with 'hidden camera' method."

Einstein remarked that a Vishniac photograph taken that day was his favorite portrait of himself.

*Pages 133, 135–141:* PHOTOGRAPHS BY ROMAN VISHNIAC, *Princeton, NJ, 1942. Published in the album* "Einstein at Work": A Suite of Seven Photographs. *Each signed and dated by the photographer.*

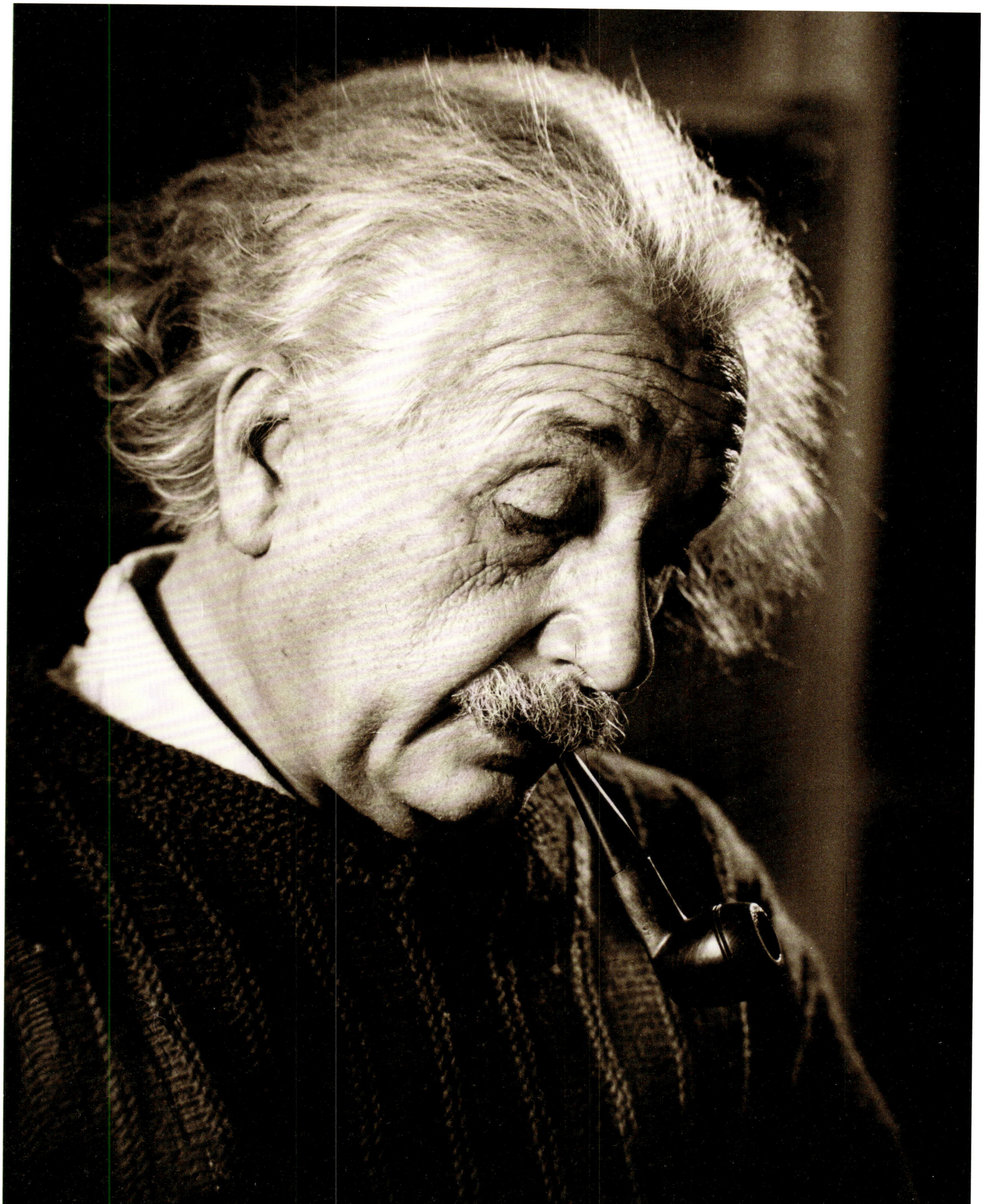

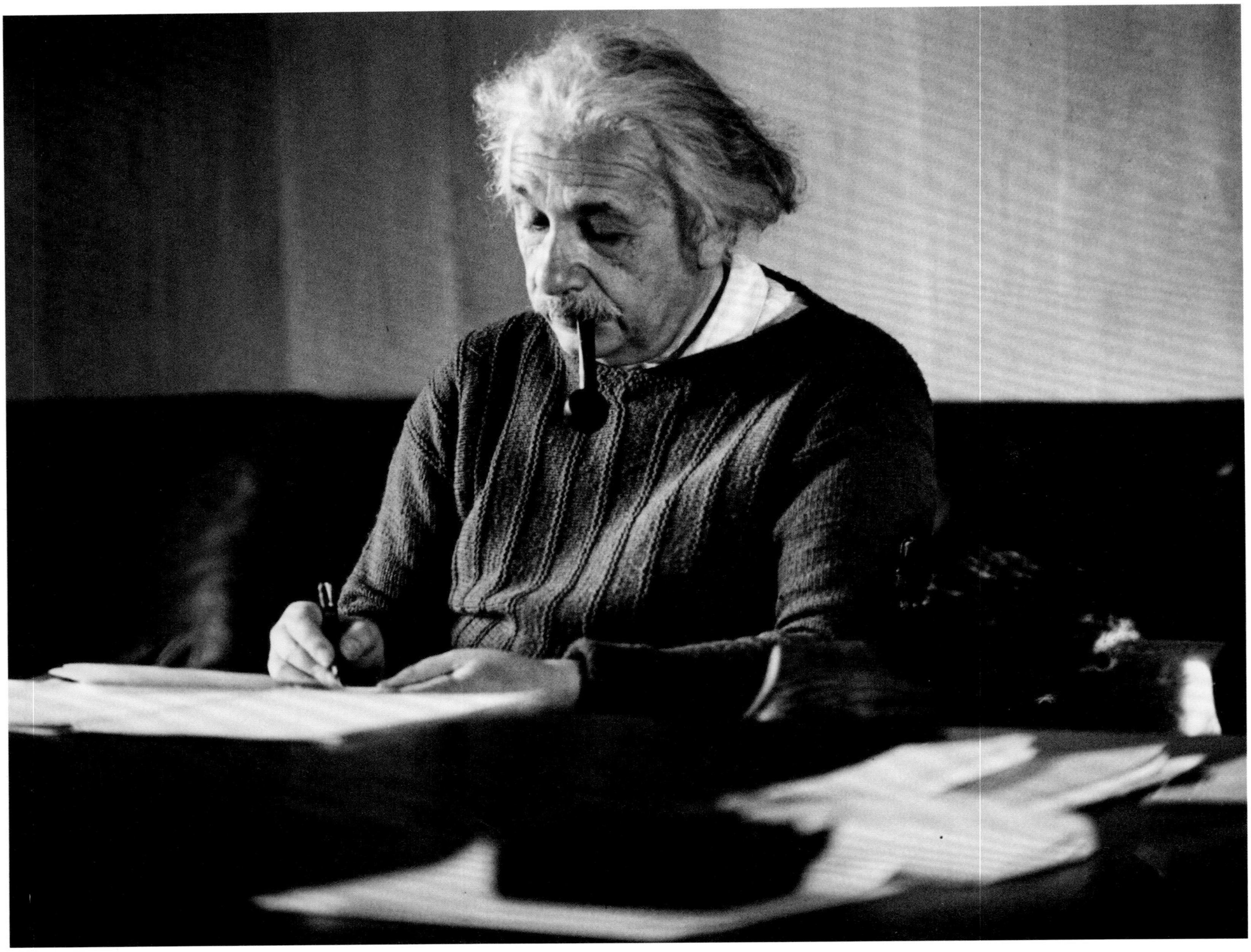

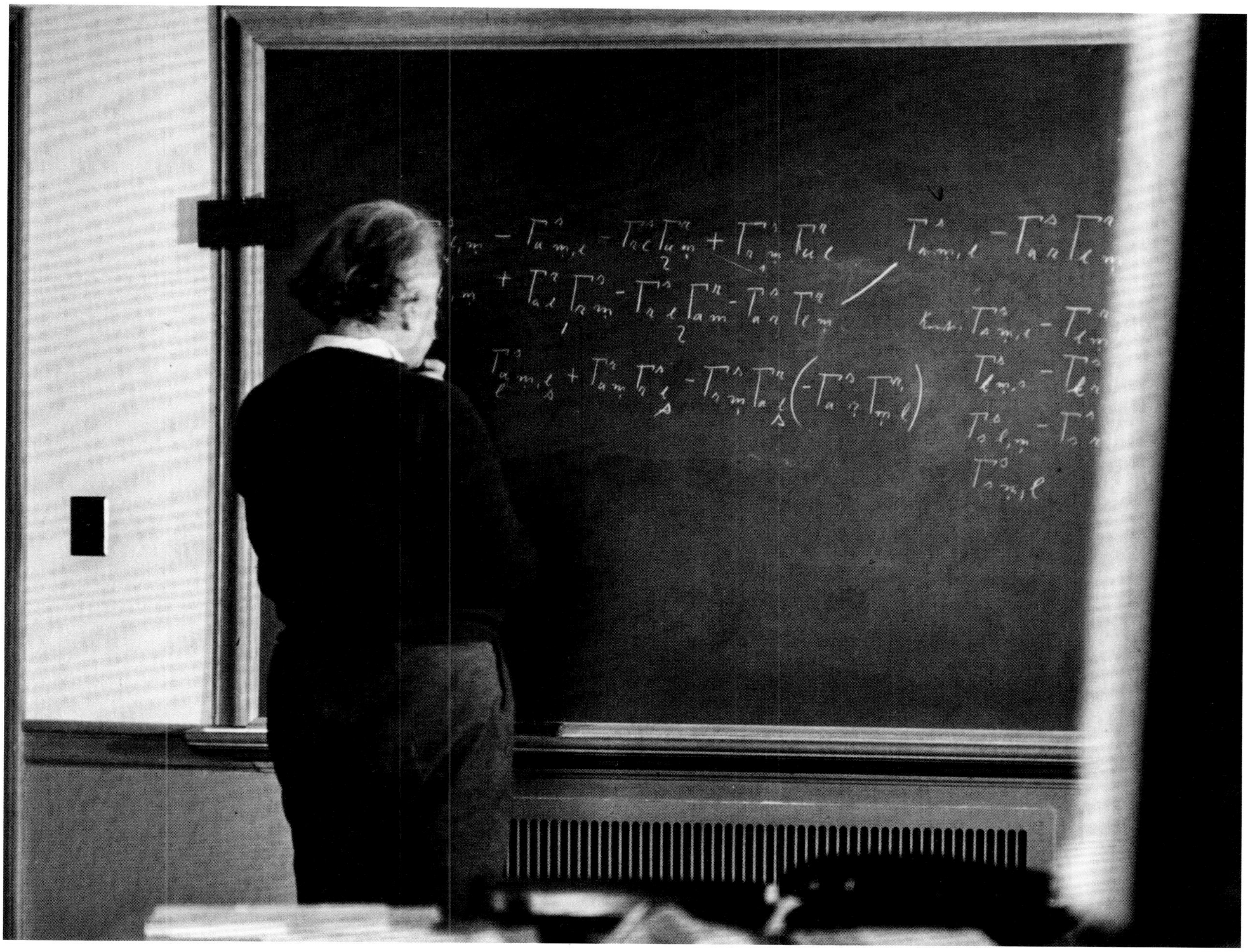

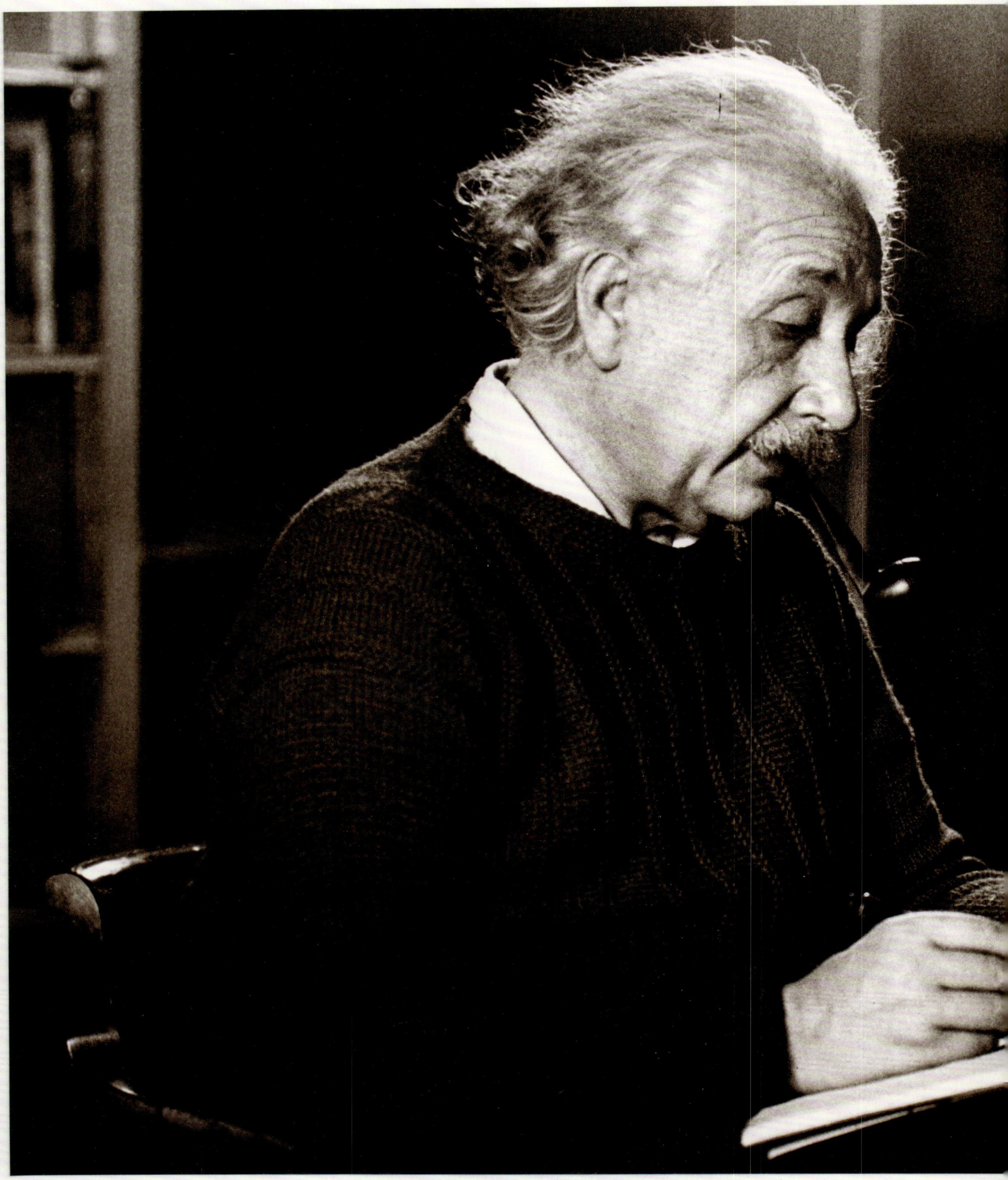

Princeton 1942.

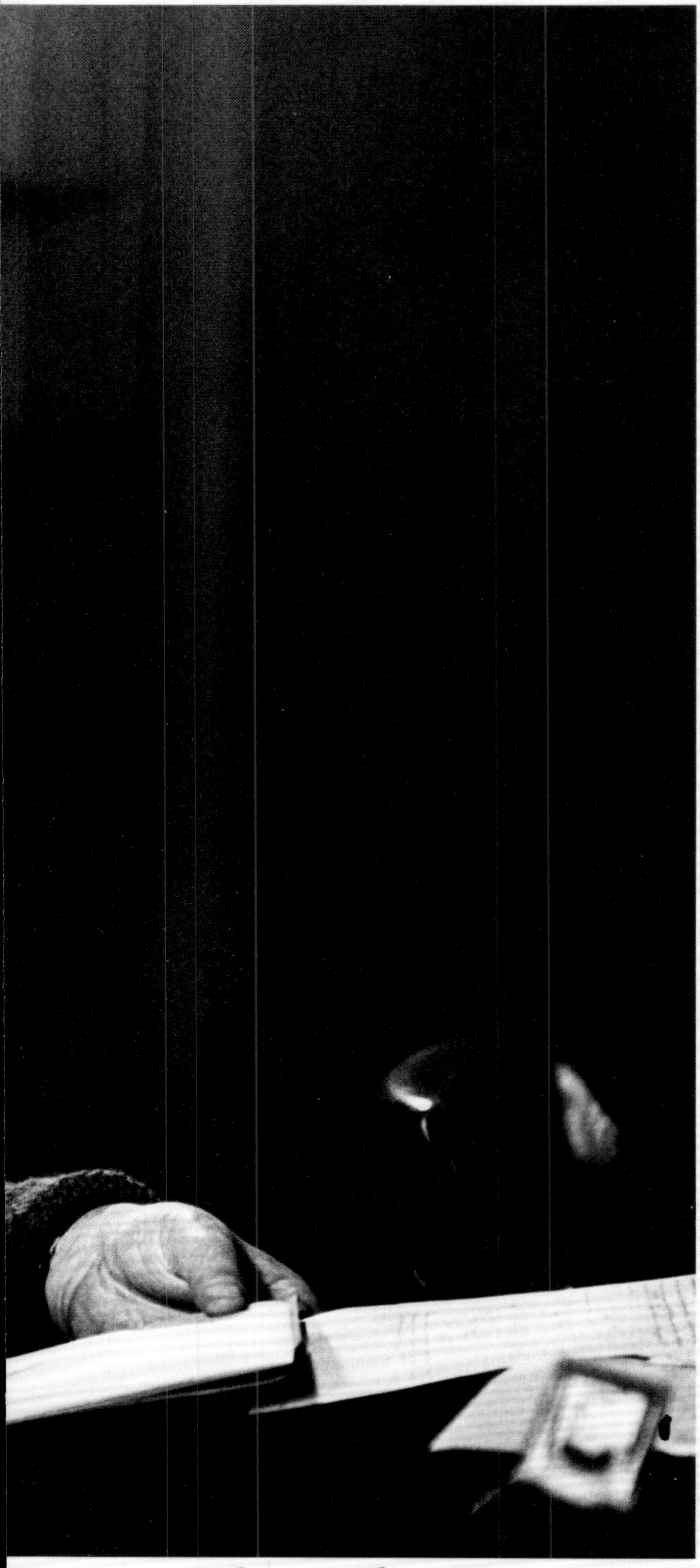

Roman Vishniac

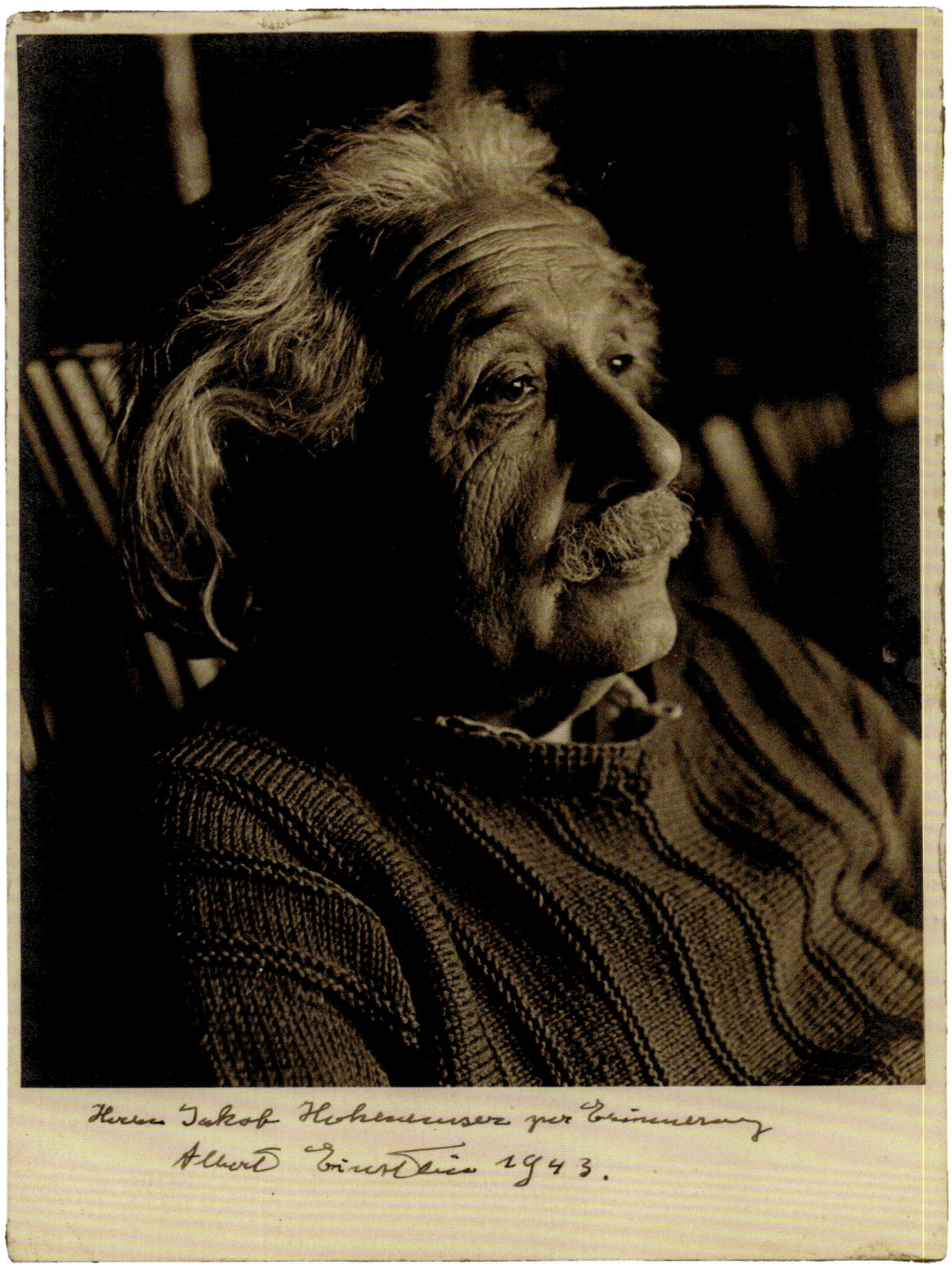
Herrn Jakob Hohenemser zur Erinnerung
Albert Einstein 1943.

### *IN HIS SIXTIES*

We get an intimate look at Einstein in his mid-sixties in this side-view bust portrait by photographer Hermann Landshoff. Einstein is seated in his study, informally dressed in a sweater, his white hair characteristically unruly. The light illuminates his thoughtful expression as he gazes into the distance.

Landshoff, like Einstein, was a German Jewish emigree who settled in the United States in the 1930s. His extensive work encompasses portraits of some of the century's most influential figures. His peers highly respected him. His work prompted the American photographer Richard Avedon to claim, "I owe everything to Landshoff."

Einstein inscribed the photo to Jakob Hohenemser, a cantor who had escaped in 1939 from the Dachau concentration camp. The inscription reads (translated from German), "As a souvenir." Einstein dated the print and signed his name in full.

PHOTOGRAPH BY HERMANN LANDSHOFF, *Princeton, NJ, 1943. Signed, dated, and inscribed by Einstein to Jakob Hohenemser: "As a souvenir."*

## *AN INFORMAL LIFE*

---

This image shows Einstein in an unusual pose, squarely facing the camera. His formal posture in front of a studio background delightfully contrasts with his attire, a leather jacket buttoned over a dress shirt.

Earlier studio portraits of Einstein typically showed him wearing a suit. This portrait of Einstein at age sixty-five reflects his more informal way of life at Princeton.

Einstein signed and dated the matte beneath this print. An accompanying letter (dated May 2, 1944), written by Gabrielle Oppenheim-Errera, the wife of Einstein's good friend Paul Oppenheim, presents this print to a friend.

Paul Oppenheim and his wife were two of Einstein's closest friends during his years in Princeton. They hosted regular Sunday luncheons for scholars at their home after Paul and Einstein took their Sunday morning walks together.

UNKNOWN PHOTOGRAPHER,
*Princeton, NJ, 1944. Signed and dated by Einstein.*

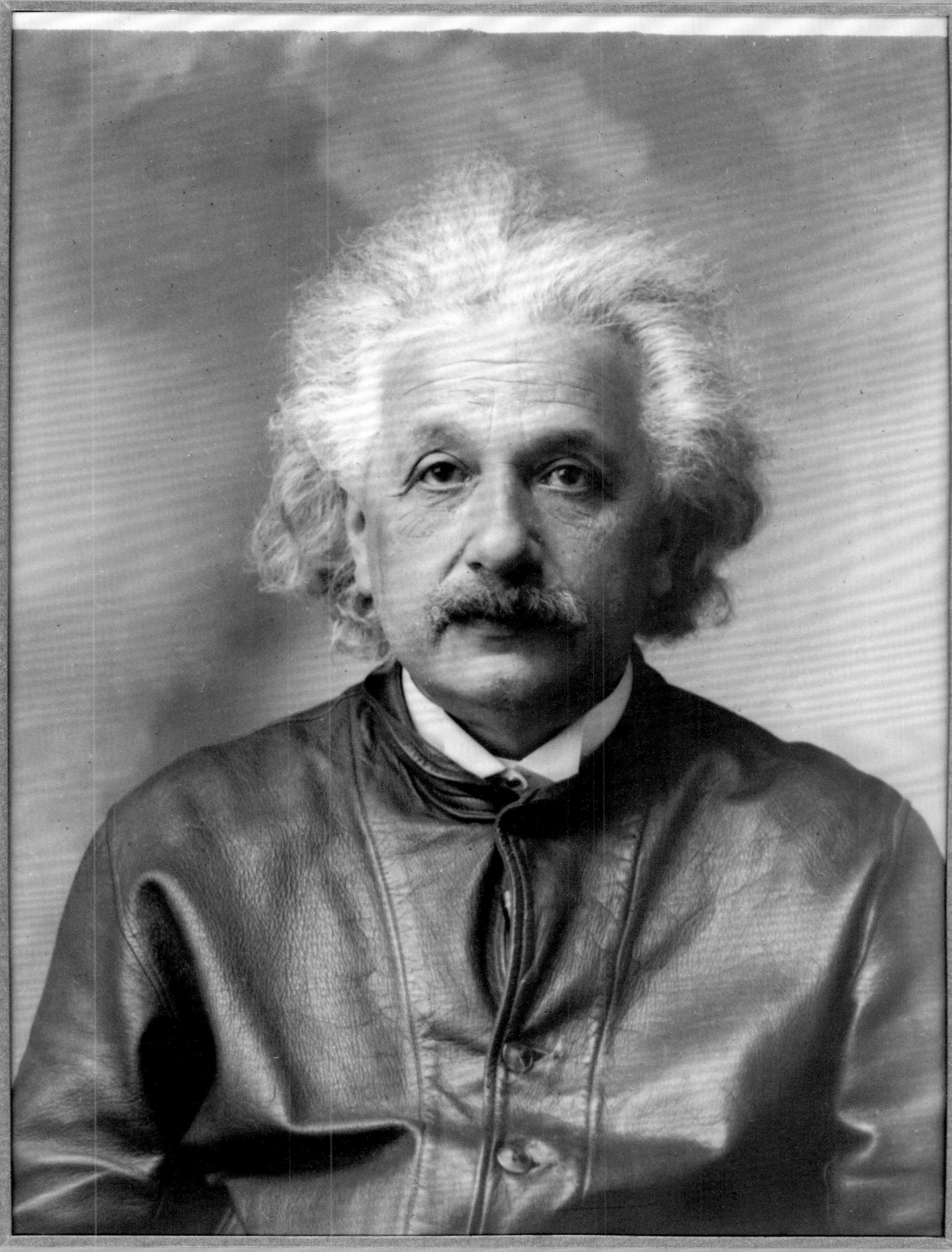

Albert Einstein 1944.

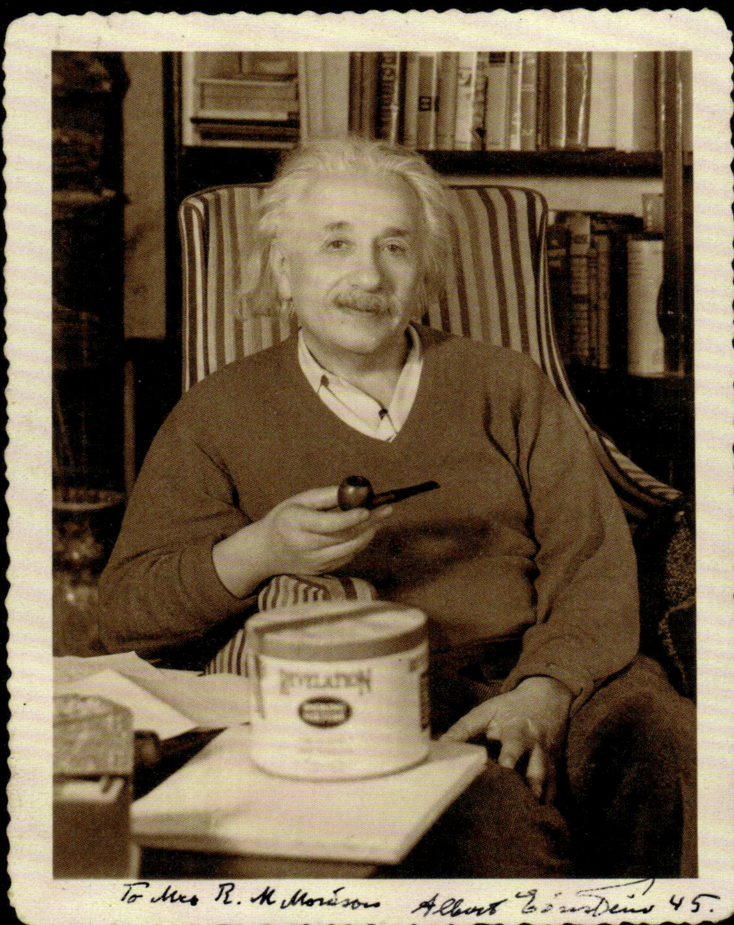
To Mrs R. M Morisson Albert Einstein 45.

### *EINSTEIN AS I KNEW HIM*

In this informal photo by Alan Windsor Richards, Einstein is relaxed and smiling in his Princeton study, pipe in hand and a large tin of tobacco in the foreground.

Richards worked as a photographer for the Palmer Physical Laboratory at Princeton University, which conducted weapons research as part of the Manhattan Project. After the war, Richards became a freelance photographer. He became close to Einstein, even publishing a book on their friendship, *Einstein as I Knew Him*.

Richards would try to spruce Einstein up a bit before their photoshoots by attempting to brush Einstein's hair behind his ears, usually to no avail. In this instance, it seems Richards had succeeded. It is one of the few photos of Einstein in his later years where his hair appears combed.

Einstein signed and inscribed the bottom margin of the print.

PHOTOGRAPH BY ALAN WINDSOR RICHARDS,
*Princeton, NJ, 1945. Signed, dated, and inscribed by Einstein: "To Mrs. R.M. Morison / Albert Einstein 45."*

*"I don't believe that the fundamental physical laws may consist in relations between probabilities for the real things, but for relations concerning the things themselves."*

—ALBERT EINSTEIN, *1945*

## *THE FUNDAMENTAL LAWS OF PHYSICS*

In this letter to British science educator Marcus Campbell Goodall, Einstein declared in a single sentence his fundamental disagreement with the state of quantum theory in 1945, writing: "I don't believe that the fundamental physical laws may consist in relations between probabilities for the real things, but for relations concerning the things themselves."

Einstein was opposed to the widely held, probability-based "Copenhagen interpretation" of quantum mechanics advocated by Niels Bohr. He objected to a theory that did not require an understanding of the properties and states of actual entities at precise moments in time. Einstein believed that such a theory could be neither a complete nor a fundamental description of nature.

CORRESPONDENCE,
*Einstein to Marcus Campbell Goodall, 1945.*

THE INSTITUTE FOR ADVANCED STUDY
SCHOOL OF MATHEMATICS
PRINCETON, NEW JERSEY

September 1o,1945

Mr.M.C.Coodall
Marconi's Wireless Telegraph Co.
Great Baddow
Essex,England



Dear Sir:

I cannot pretend to have grasped the ideas which you have indicated in your letter of August 12th. The reason may be that my own expectations concerning the future basis of physics are very different from your own. This is because I don't believe that the fundamental physical laws may consist in relations between probabilities for the real things, but for relations concerning the things themselves (fields f.i.). For this reason I am not enough aquainted with the attempts to expand the contemporary quantum-mechanics to fields and I am not able to give judgement about such efforts. I know that my friend,Professor Max Born (University of Edinburgh) has worked arduously in this direction so that I suggest to you to get in touch with him about the matter.

Yours very sincerely,

A. Einstein.

Albert Einstein.

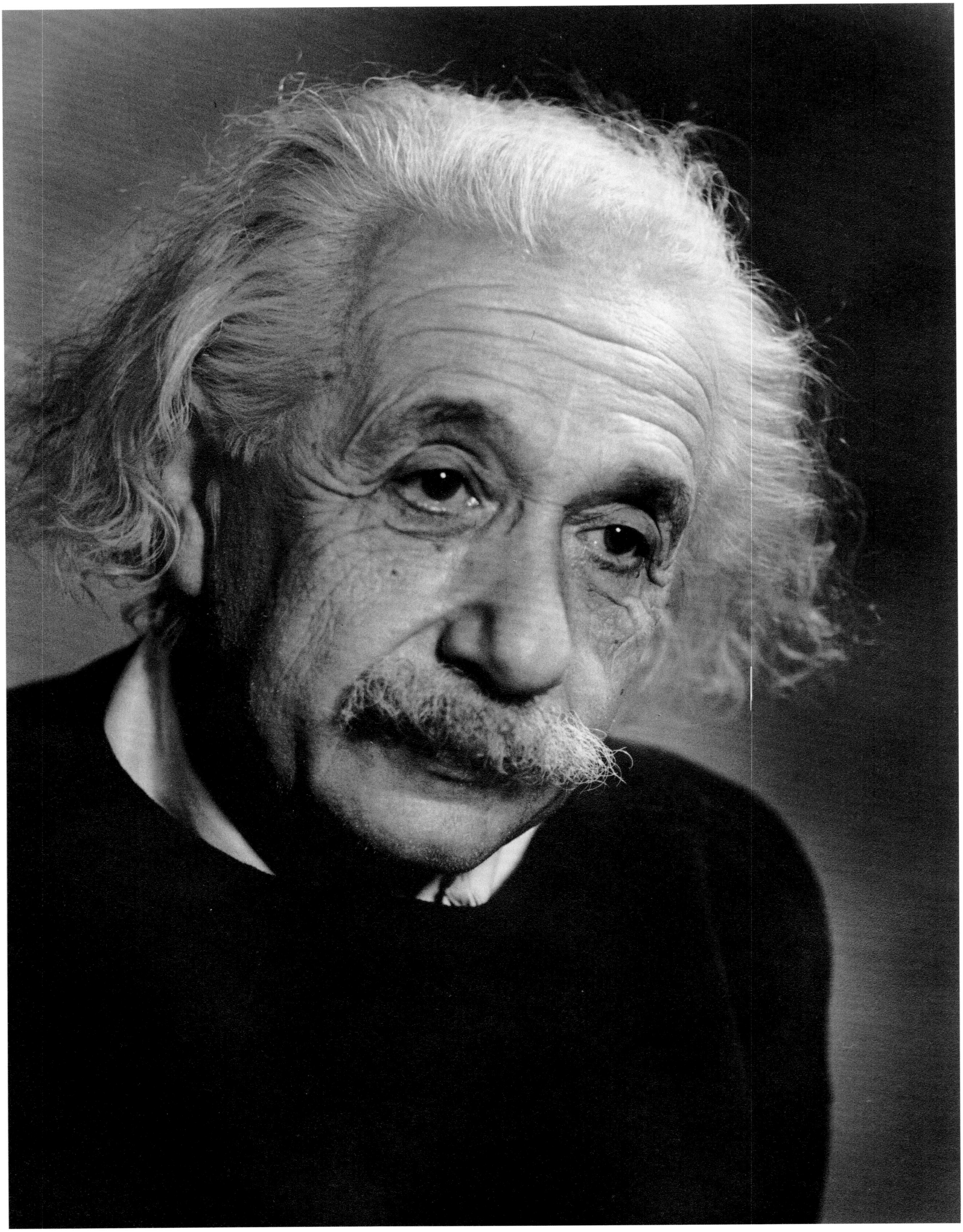

## *TEN MINUTES OF TIME*

"In 1946, when Albert Einstein was in residence at Princeton University, Fred Stein was granted ten minutes of the great man's time to take a portrait. After the time was up, Einstein's secretary came in to usher Stein out. However, Einstein insisted that he stay, saying that their discussion was too interesting to cut short. The secretary came back repeatedly, but the visit extended to two hours. The resulting portrait by Fred Stein pictures a deep intelligence engaged in thought. It became an iconic image, and one of the most famous photographs ever taken of Albert Einstein."

Like many other Einstein photographers, Stein had escaped from Nazi Germany. He was a highly regarded portrait photographer, producing celebrated images of some of the most influential people of his time, including Georgia O'Keeffe, Frank Lloyd Wright, Hannah Arendt, and Helen Keller. This portrait of Einstein is his most famous.

This print is one of 450 produced after 1967 on behalf of the Fred Stein estate by the photographer's son, Peter Stein. It was "printed and archivally processed by a master printer under Peter Stein's supervision and to his approval, and matched as closely as possible to his father's vintage prints."

PHOTOGRAPH BY FRED STEIN,
*Princeton, NJ, 1946.*

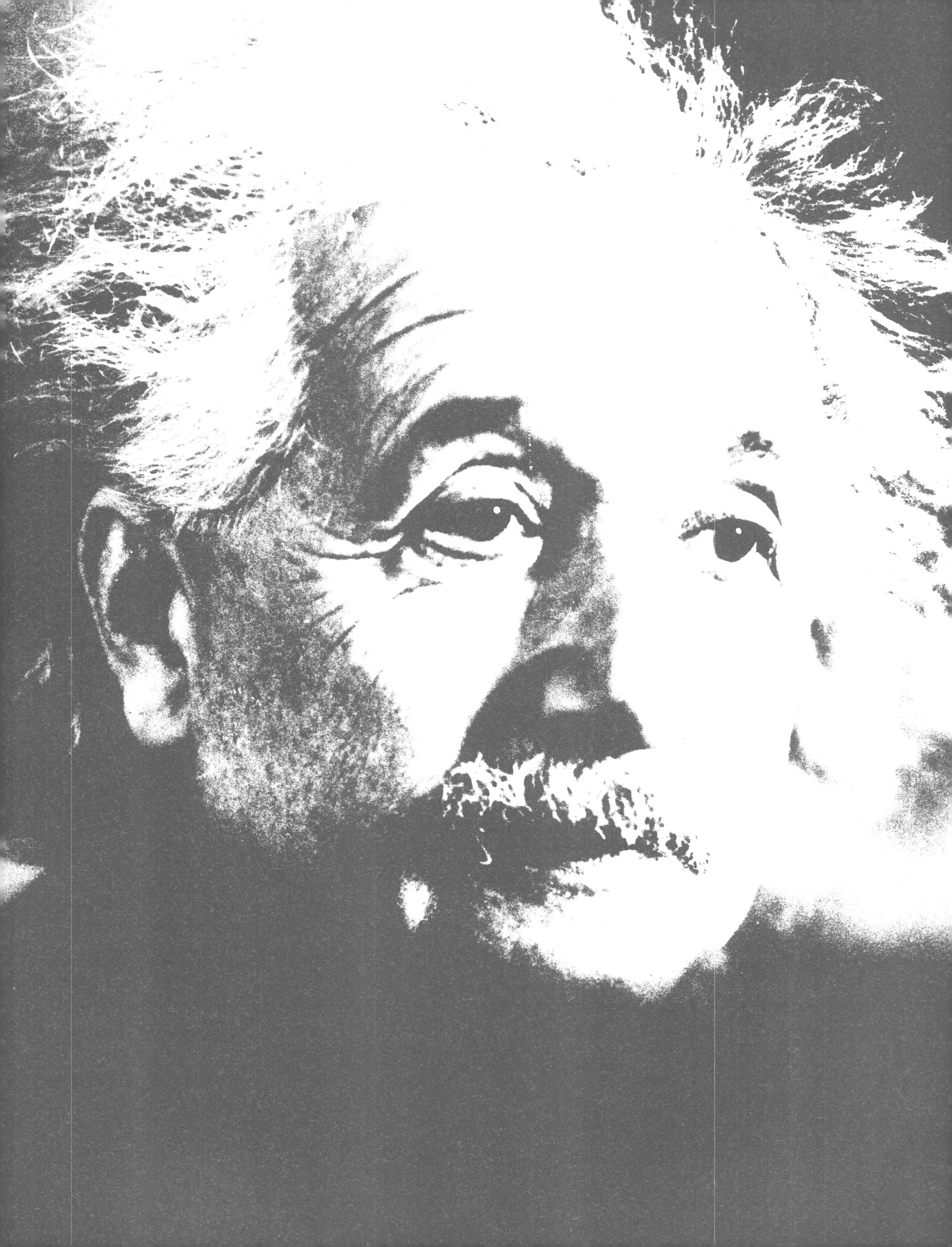

*The eternal mystery of the world is its comprehensibility. . . . The fact that it is comprehensible is a miracle."*

—ALBERT EINSTEIN, *1936*

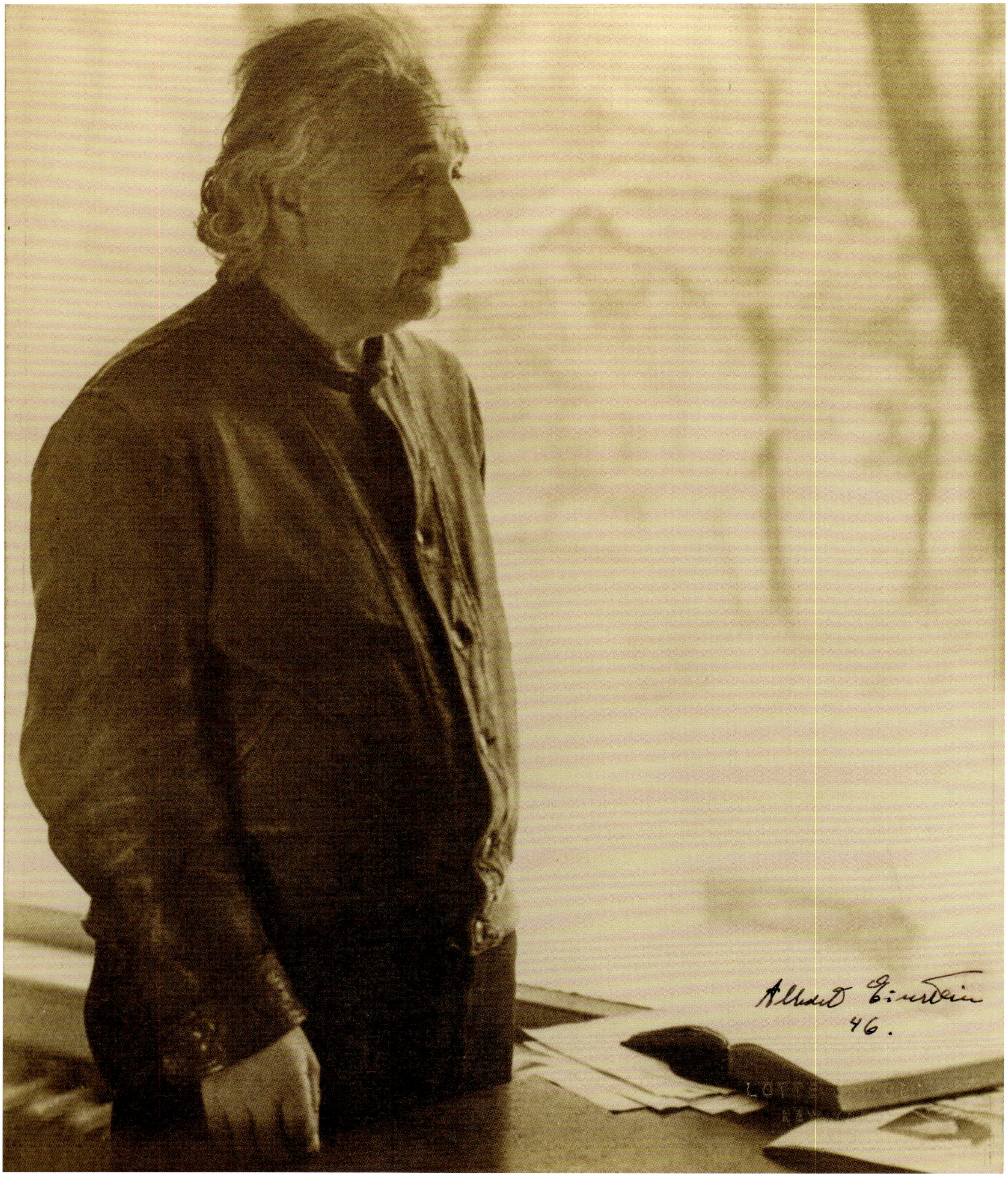
Albert Einstein
46.

## *THINKING*

This unusual sepia-toned photograph by Lotte Jacobi shows Einstein standing in his office at the Institute for Advanced Study, apparently deep in thought. As in the past, Jacobi captured a less formal Einstein, away from a studio and at ease in his beloved leather jacket. Light from the window sets Einstein's figure in dramatic relief in this artful composition.

In this image, as in many other postwar photographs, Einstein's expression conveys a bit of sadness, perhaps influenced by recent world events.

Einstein signed in full and dated this print in dark ink. Its bottom right bears Jacobi's two-line blind stamp.

PHOTOGRAPH BY LOTTE JACOBI,
*Princeton, NJ, 1946. Signed and dated by Einstein.*

## *AMERICAN IMMIGRANTS*

"Trude Fleischmann, who developed a passion for photography already as a child, rapidly became one of Vienna's leading portrait photographers soon after opening her own studio at the age of twenty-five. . . . Her outstanding portraits of intellectuals and artists . . . remain an important record of twentieth-century European culture. Because of her Jewish background, Fleischmann was forced to seek work elsewhere after the Anschluss in 1938. Leaving behind most of her negatives, she emigrated to Paris, London and eventually . . . to New York," where she opened up a studio in 1940. Once in the United States, Fleischmann sought out influential people for her subjects, focusing on American immigrants.

Fleischmann took this rare and dramatic photograph of Einstein against a dark background. She signed and dated the matte at the bottom left in 1946. Five years later, Einstein added his signature.

PHOTOGRAPH BY TRUDE FLEISCHMANN, *New York, 1946.*
*Signed and dated by Einstein and the photographer.*

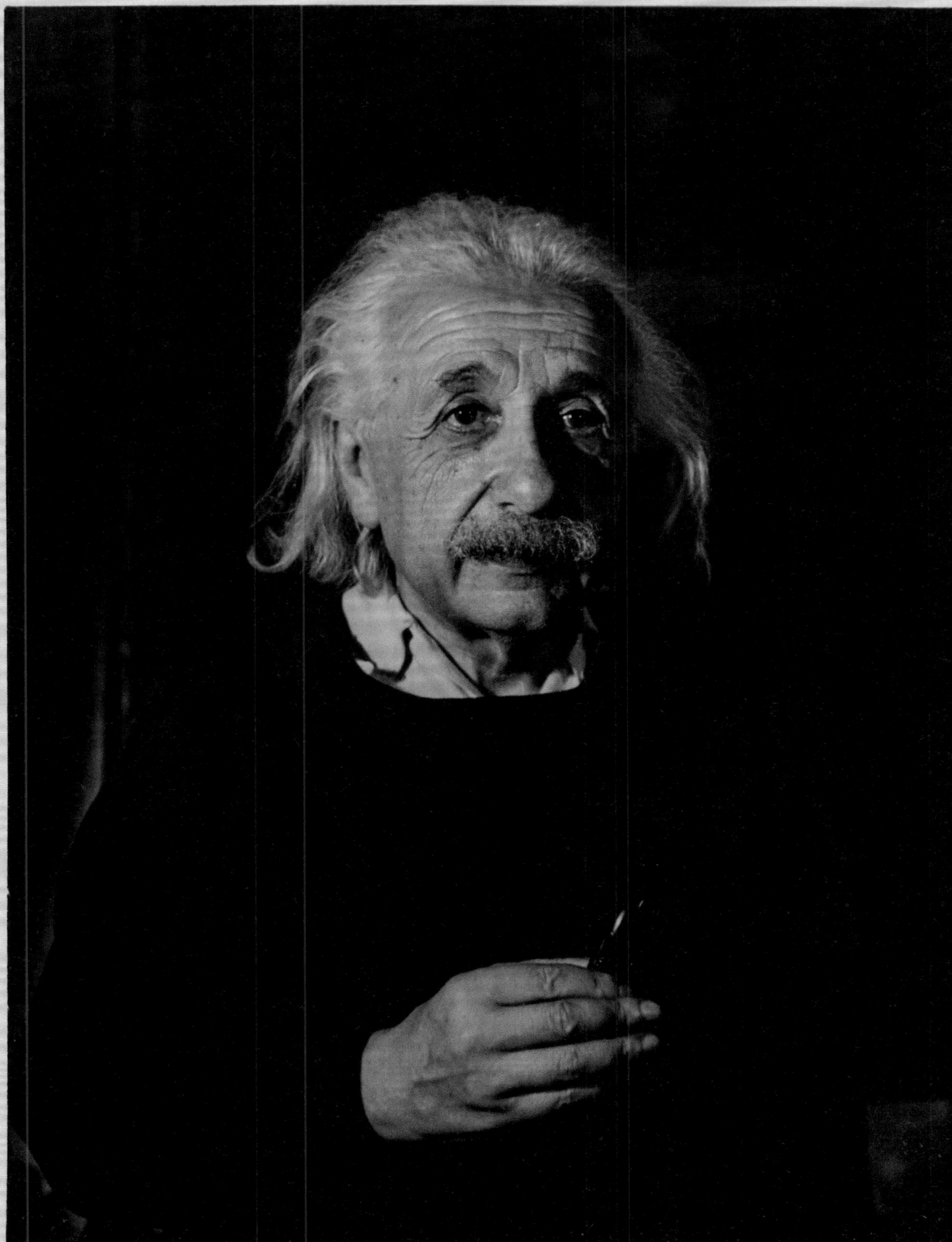

Trude Fleischmann
1946

A. Einstein. 51.

## *PERSON OF THE CENTURY*

This iconic photograph of Einstein by Philippe Halsman has become one of the most recognizable images of the twentieth century. It appeared on a 1966 US postage stamp and was featured on the cover of the December 31, 1999, edition of *Time* magazine, which honored Einstein as the "Person of the Century." The example of the famous photograph shown here is a rare lifetime print, signed by Halsman in the bottom-left corner. In his book *Philippe Halsman: A Retrospective*, Halsman explained the circumstances of the photo:

"I admired Albert Einstein more than anyone I ever photographed, not only as the genius who single-handedly had changed the foundation of modern physics but even more as a rare and idealistic human being. Personally, I owed him an immense debt of gratitude. After the fall of France, it was through his personal intervention that my name was added to the list of artists and scientists who, in danger of being captured by the Nazis, were given emergency visas to the United States. After my miraculous rescue I went to Princeton to thank Einstein, and I remember vividly my first impression. Instead of a frail scientist I saw a deep-chested man with a resonant voice and a hearty laugh. . . . The question of how to capture the essence of such a man in a portrait filled me with apprehension. Finally, in 1947, I had the courage to bring on one of my visits my Halsman camera and a few floodlights. After tea, I asked for permission to set up my lights in Einstein's study. The professor sat down and started peacefully working on his mathematical calculations. I took a few pictures. Ordinarily, Einstein did not like photographers, whom he called *Lichtaffen* (light monkeys). But he cooperated because I was his guest and, after all, he had helped save me. Suddenly looking into my camera, he started talking. He spoke about his despair that his formula $E=mc^2$ and his letter to President Roosevelt had made the atomic bomb possible, that his scientific research had resulted in the death of so many human beings. 'Have you read,' he asked, 'that powerful voices in the United States are demanding that the bomb be dropped on Russia now, before the Russians have time to perfect their own?' With my entire being I felt how much this infinitely good and compassionate man was suffering from the knowledge that he had helped to put in the hands of politicians a monstrous weapon of devastation and death. He grew silent. His eyes had a look of immense sadness. There was a question and a reproach in them. The spell of this moment almost paralyzed me. Then, with an effort, I released the shutter of my camera. Einstein looked up, and I asked him, 'So you don't believe that there will ever be peace?' 'No,' he answered. 'As long as there will be man there will be wars.'"

PHOTOGRAPH BY PHILIPPE HALSMAN,
*Princeton, NJ, 1947. Signed by the photographer.*

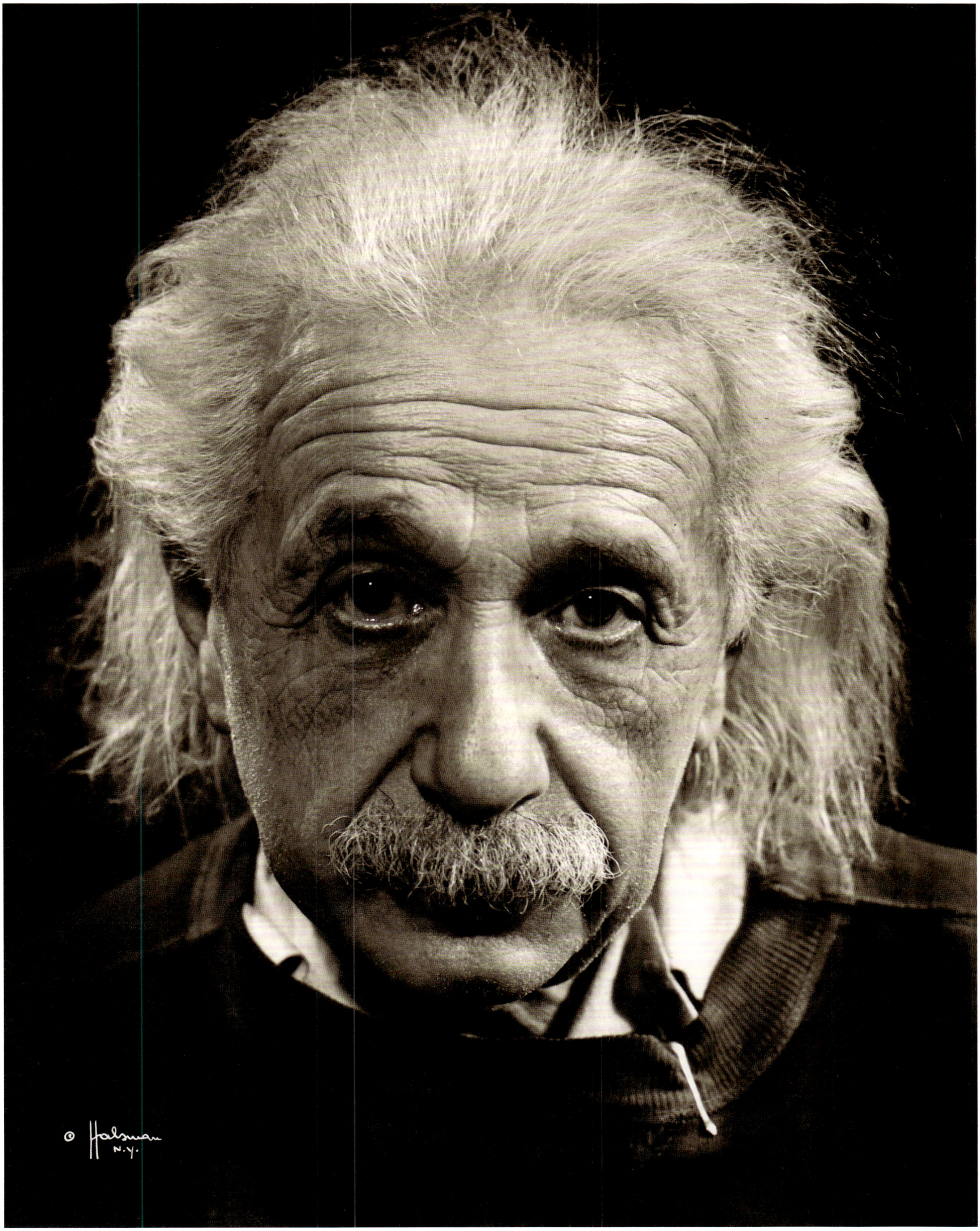
© Halsman
N.Y.

## *THE GRACE OF CLARITY*

Einstein sat for photographer Yousuf Karsh in 1948. Karsh had survived the Armenian genocide and was a refugee who eventually became a Canadian citizen. He was recognized as one of the greatest working portrait photographers. Karsh's photographs are among the most famous images of Einstein.

This meticulously lit image was chosen as the frontispiece for the book *Albert Einstein, Philosopher-Scientist* (1949).

Einstein signed and dated the print in 1953, inscribing in German a poetic thought that captures both the moment of scientific insight and the quality of this photograph. He had difficulty writing on the photo's glossy surface, attempting first to write on the left side of the bottom margin before inscribing it successfully on the right side. The inscription (translated from the original German) reads: "The grace of clarity often comes to you, in uniting precision and depth."

PHOTOGRAPH BY YOUSUF KARSH, *1948.*
*Signed, dated, and inscribed by Einstein in 1953.*

*The grace
of clarity often
comes to you,
in uniting precision
and depth."*

—ALBERT EINSTEIN, *1948*

“The grace
of clarity often
comes to you,
in writing precision
and depth.”

—ALBERT EINSTEIN, 1948

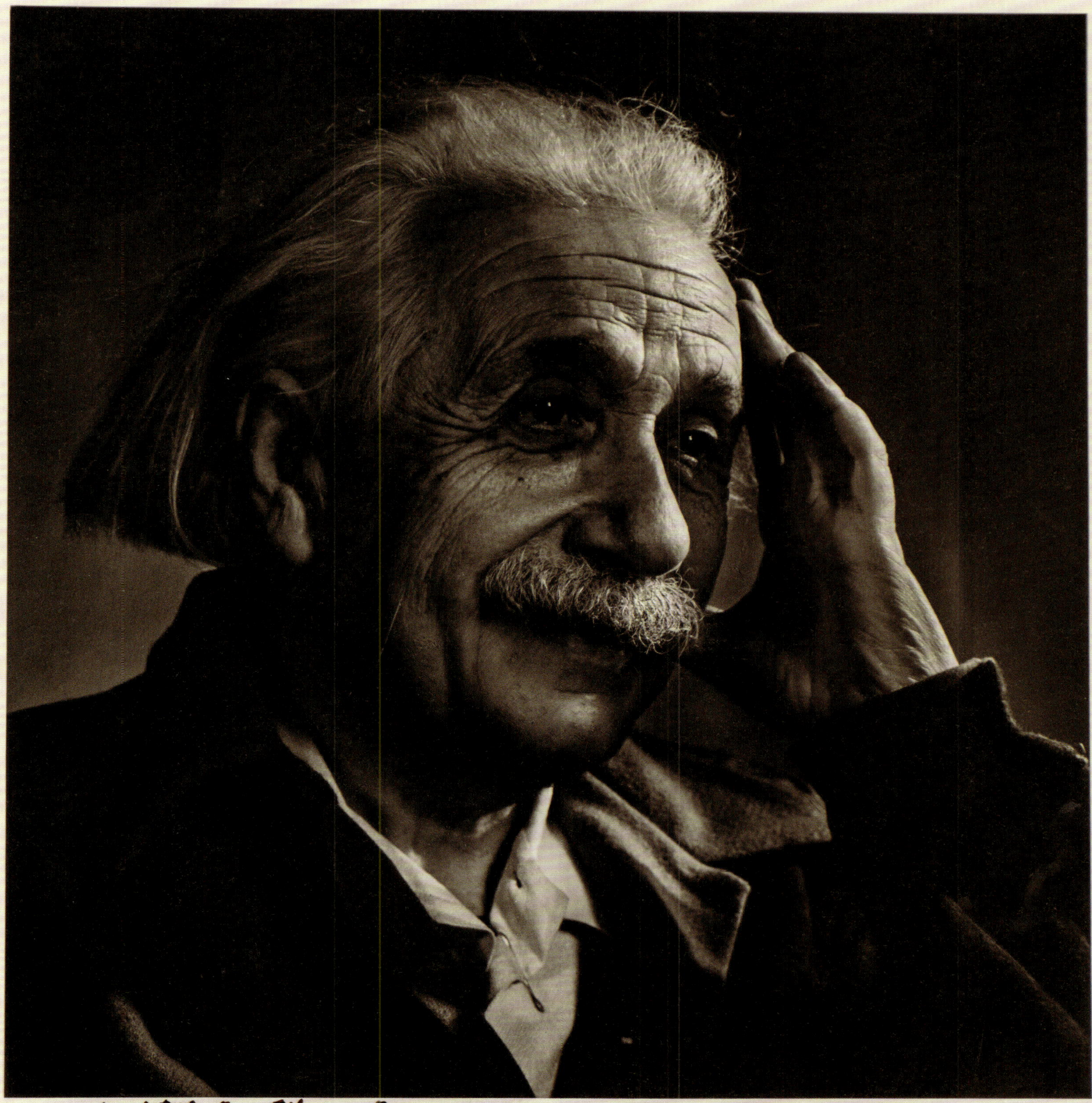

Oft [illegible] Kürze und Tiefe zu [illegible]. A. Einstein. 53 Oft ward dir die Gnade
Klarheit, Kürze und Tiefe zu vereinen.

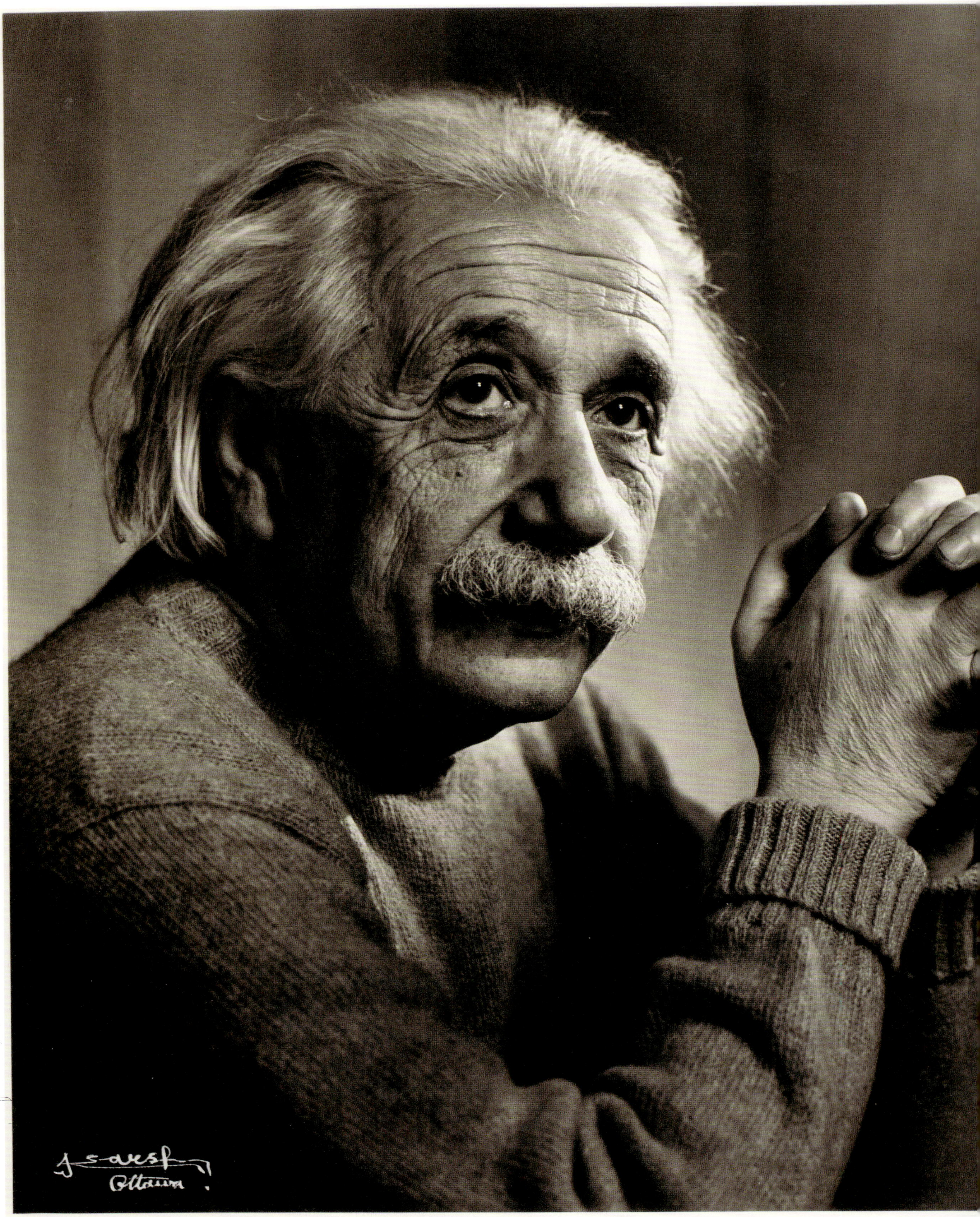

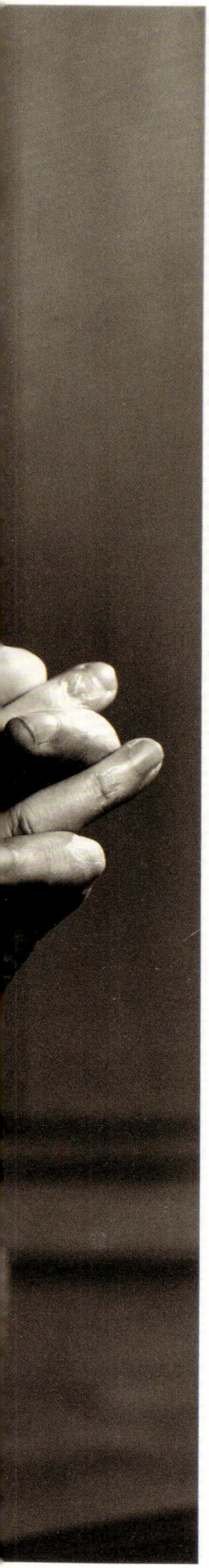

## *ROUGH GRANDEUR*

Yousuf Karsh was powerfully drawn to Einstein, as he explained in his book *Regarding Heroes*: "Among the tasks that life as a photographer had set me, a portrait of Albert Einstein had always seemed a 'must'—not only because this greatest refugee of our century has been accounted by all the world as the [most] outstanding scientist since Newton, but because his face, in all its rough grandeur, invited and challenged the camera."

This masterful photograph of a serene Einstein with hands clasped, almost as if in prayer, is among the most recognizable portraits of him in his later years. The print is elegantly signed in white by Karsh.

PHOTOGRAPH BY YOUSUF KARSH,
*1948. Signed by the photographer.*

*Although I am a typical
loner in daily life,
my consciousness of
belonging to the invisible
community of those who
strive for truth, beauty,
and justice has preserved
me from feeling isolated."*

—ALBERT EINSTEIN, *1932*

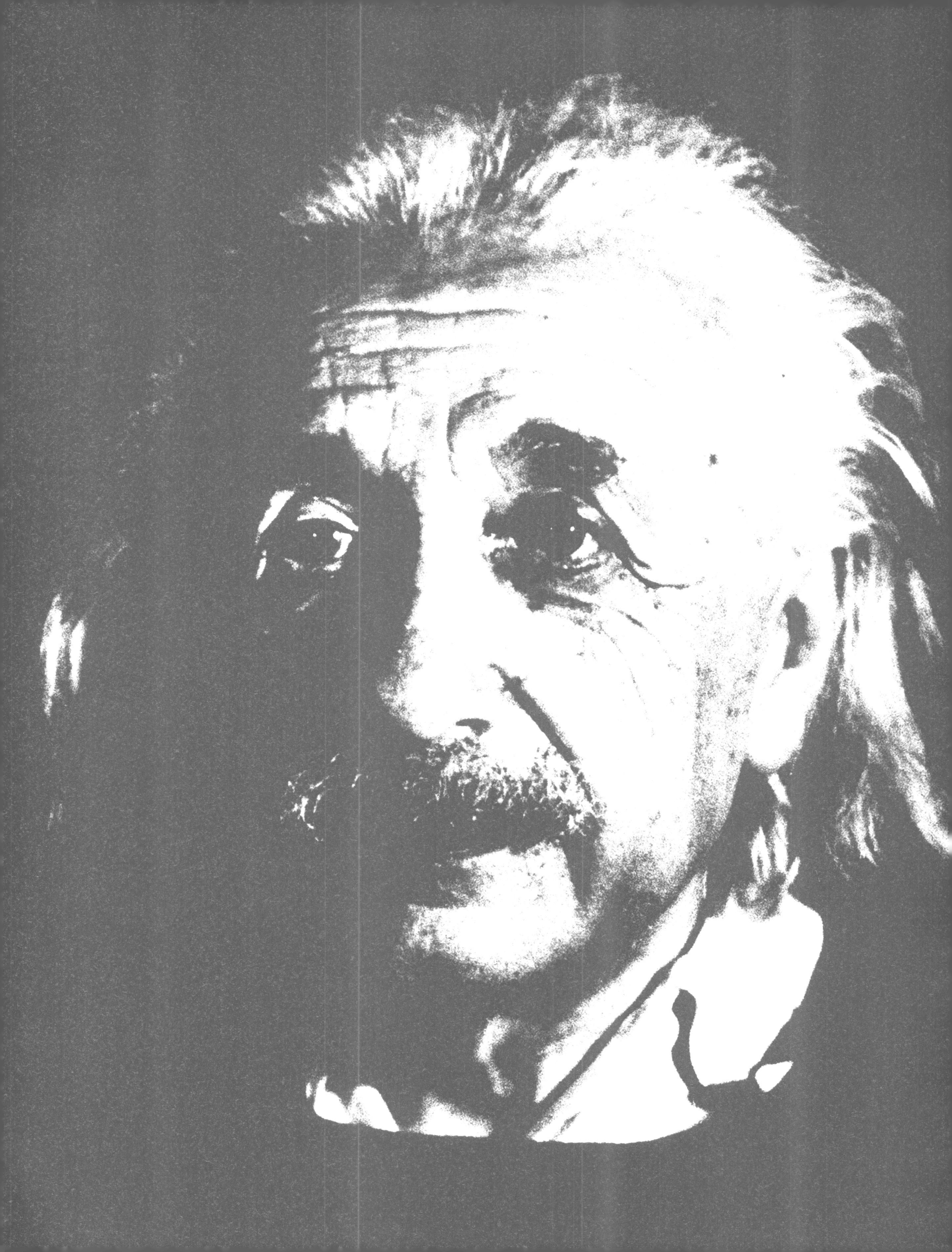

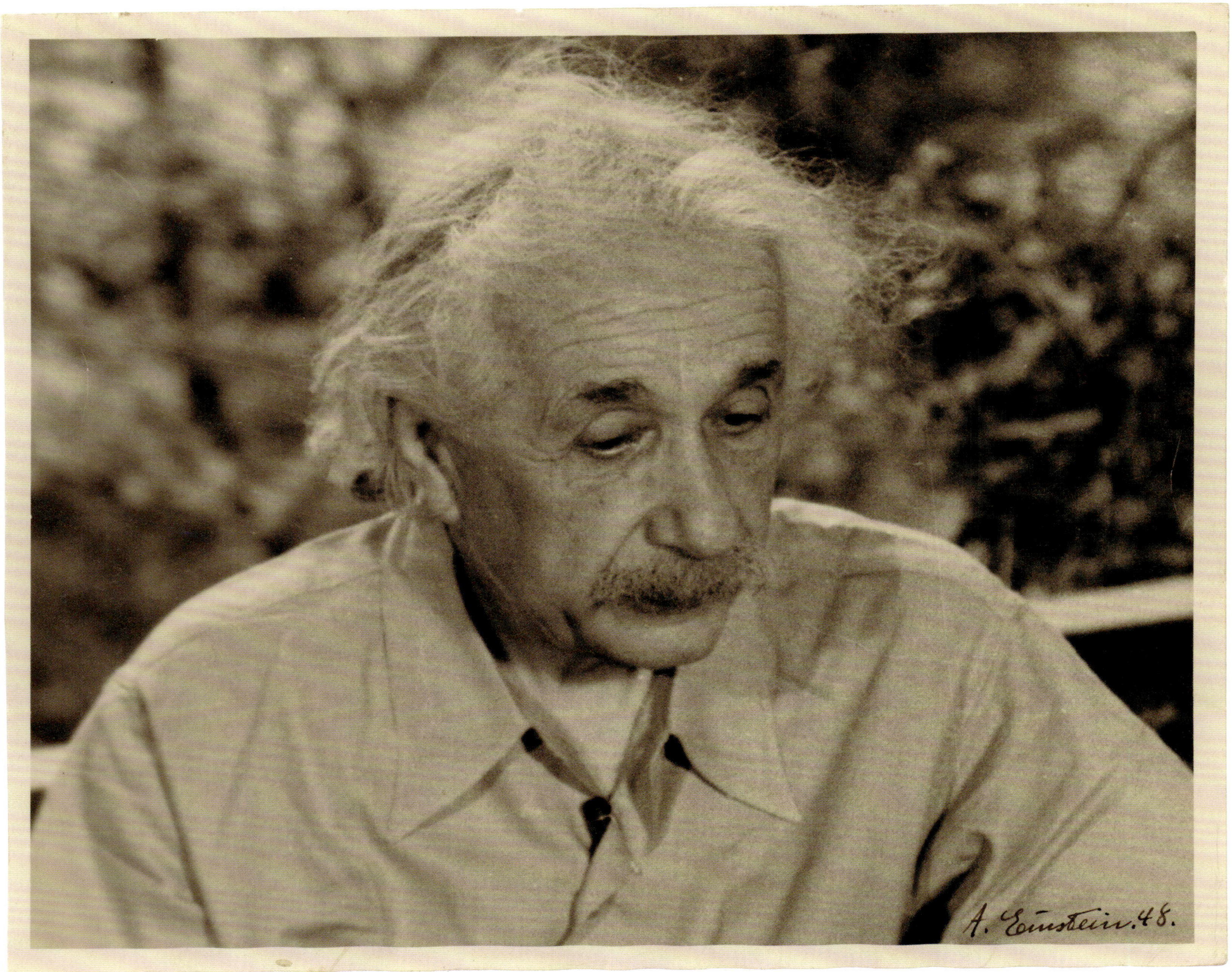
A. Einstein. 48.

## *ATOMIC POWER*

To celebrate the end of the Second World War, in August 1946 the newsreel service The March of Time released *Atomic Power*, a short film portraying the events leading up to the atomic bomb's creation. The film featured scientists involved in the war effort reenacting historical moments.

This photo, signed by Einstein in 1948, was taken during the filming of the movie when he was asked to re-create (with physicist Leó Szilárd) the drafting of the famous 1939 "Einstein-Szilárd letter" to President Roosevelt. The letter alerted Roosevelt to the scientific potential to create "extremely powerful bombs of a new type." One of the most influential documents of the twentieth century, the letter spurred Roosevelt to action, leading to the Manhattan Project and the development of the atomic bomb.

Einstein did not work on the bomb himself. After the war, he expressed his regret about having signed the letter. In an interview with *Newsweek* in 1947, he reflected, "Had I known that the Germans would not succeed in developing an atomic bomb, I would have done nothing."

UNKNOWN PHOTOGRAPHER,
*1948. Signed and dated by Einstein.*

## *LIGHTNESS OF BEING*

---

This image of Einstein, age seventy, was taken by Alan Windsor Richards in 1949. It is quite different from Richards's 1945 photograph of Einstein relaxing in his study. Here, Einstein is standing, the camera below and angled upward. Illuminated by a bright light, his face has a rather stern expression. It is an unusual image, in contrast to the more common portrayals of Einstein in this period as a kindly, thoughtful sage.

The bottom margin of the matte is signed and dated by Einstein and signed by Richards.

PHOTOGRAPH BY ALAN WINDSOR RICHARDS, *1949. Signed and dated by Einstein and signed by the photographer.*

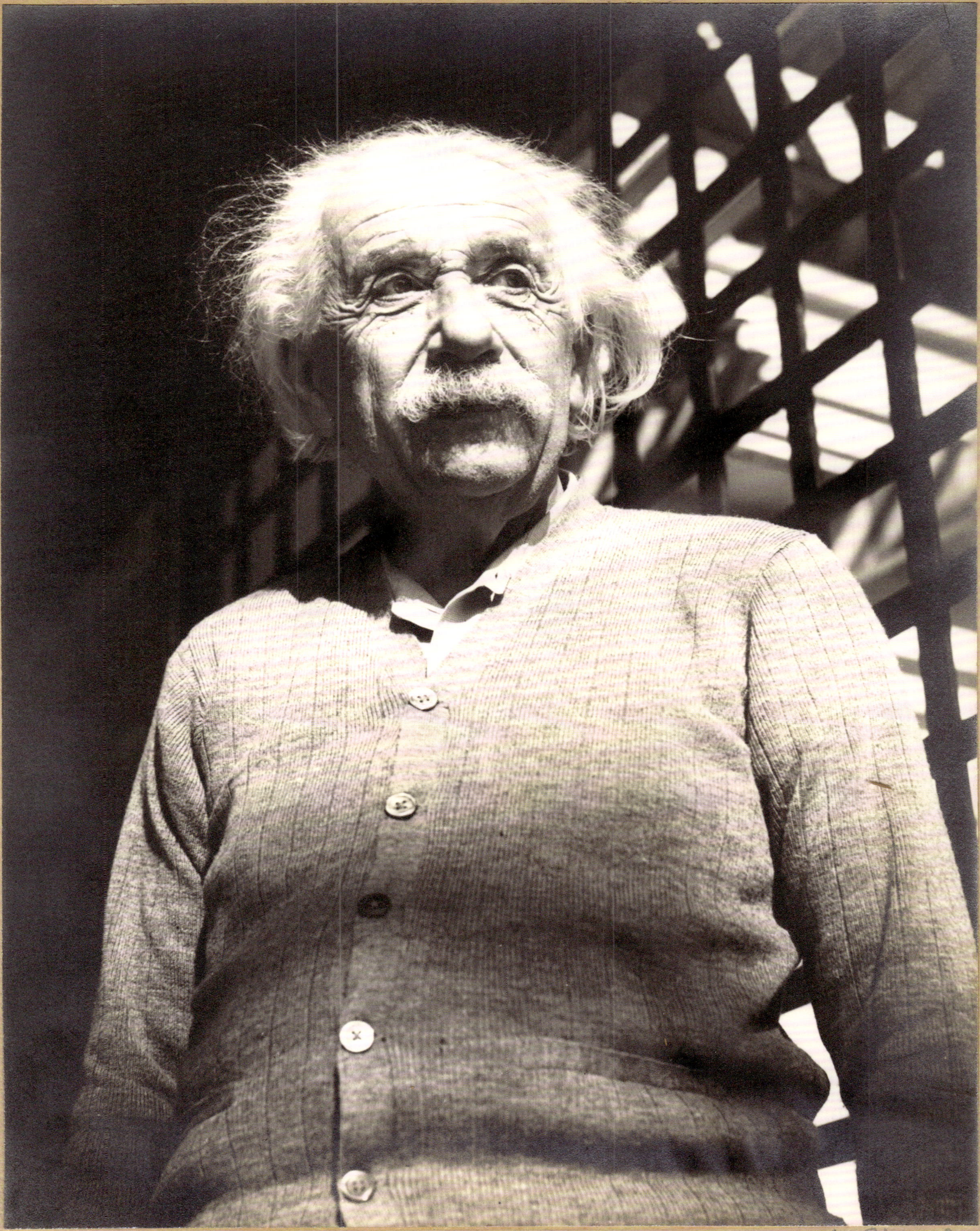

A. Einstein. 49.

A W Richards

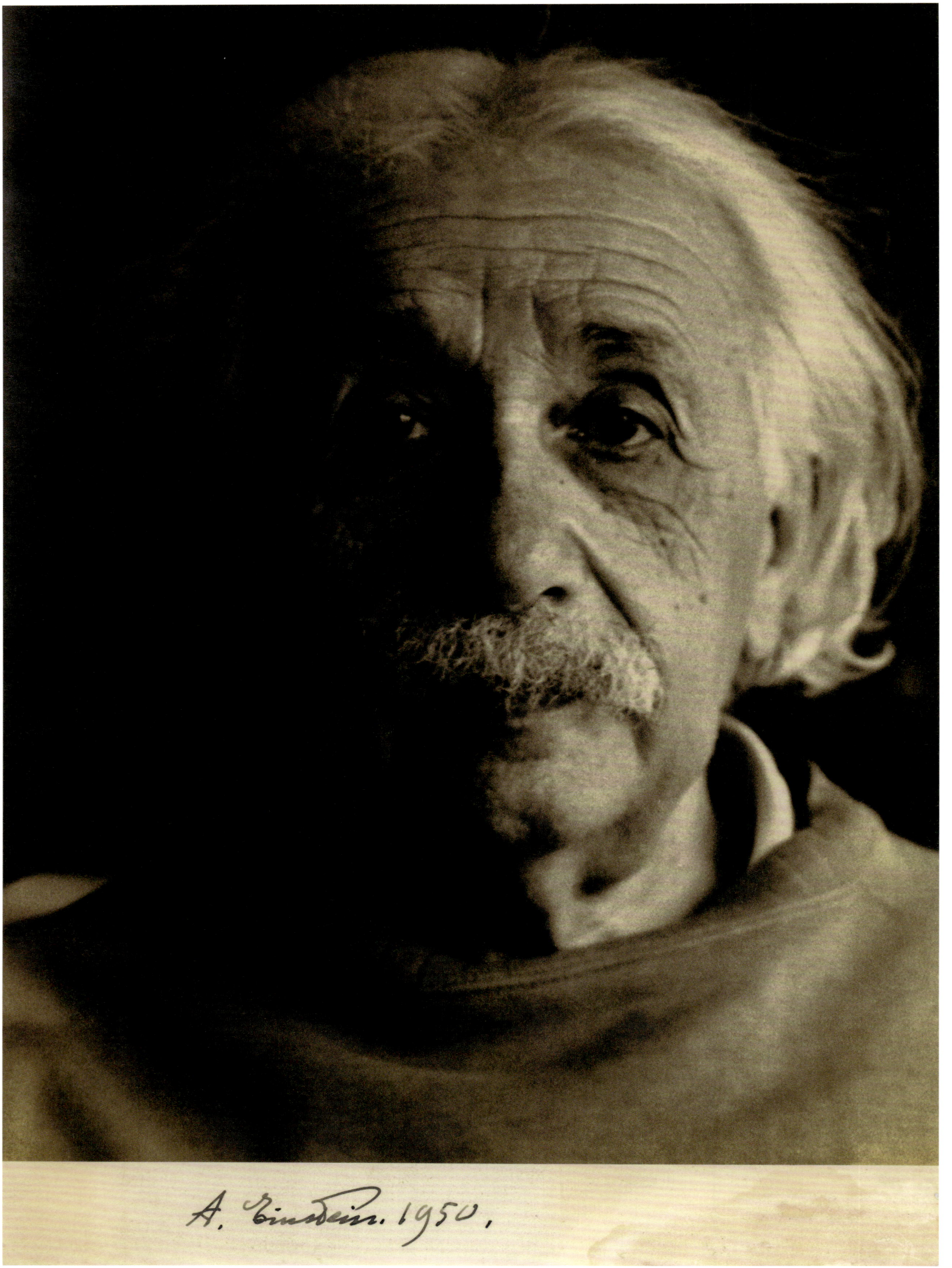
A. Einstein. 1950.

## *AT HOME IN PRINCETON*

Hermann Landshoff visited Einstein and took photographs several times in the 1940s and early 1950s. He amassed a beautiful body of work showing Einstein in quieter moments at home.

This powerful photo, signed by Einstein in 1950, is an excellent example of the communicative power of Einstein's face, particularly his eyes. Although the composition is severe—a close-up of the subject's face staring straight into the camera, half-bathed in shadow—Einstein's expression is soft and tender.

PHOTOGRAPH BY HERMANN LANDSHOFF, *Princeton, NJ, 1950. Signed and dated by Einstein.*

### *SIGNS OF AGING*

This oversized photo by the Hungarian American photographer Marcel Sternberger commands the viewer's attention. Despite the lines and wrinkles of aging, Einstein's eyes are as bright and energetic as in his younger days.

Sternberger created celebrated images of some of the most influential people of his time, including Franklin Delano Roosevelt and Sigmund Freud.

Einstein found Sternberger's photographs to be particularly impressive, writing in a letter to Sternberger: "The portraits you have made of me are true masterpieces of the photographic art. It seems quite amazing to me that you could present this subject so appetizingly."

The print—dated and inscribed by Einstein to Sternberger—was the photographer's personal copy.

PHOTOGRAPH BY MARCEL STERNBERGER, *Princeton, NJ, 1950. Signed, dated, and inscribed by Einstein and signed by the photographer.*

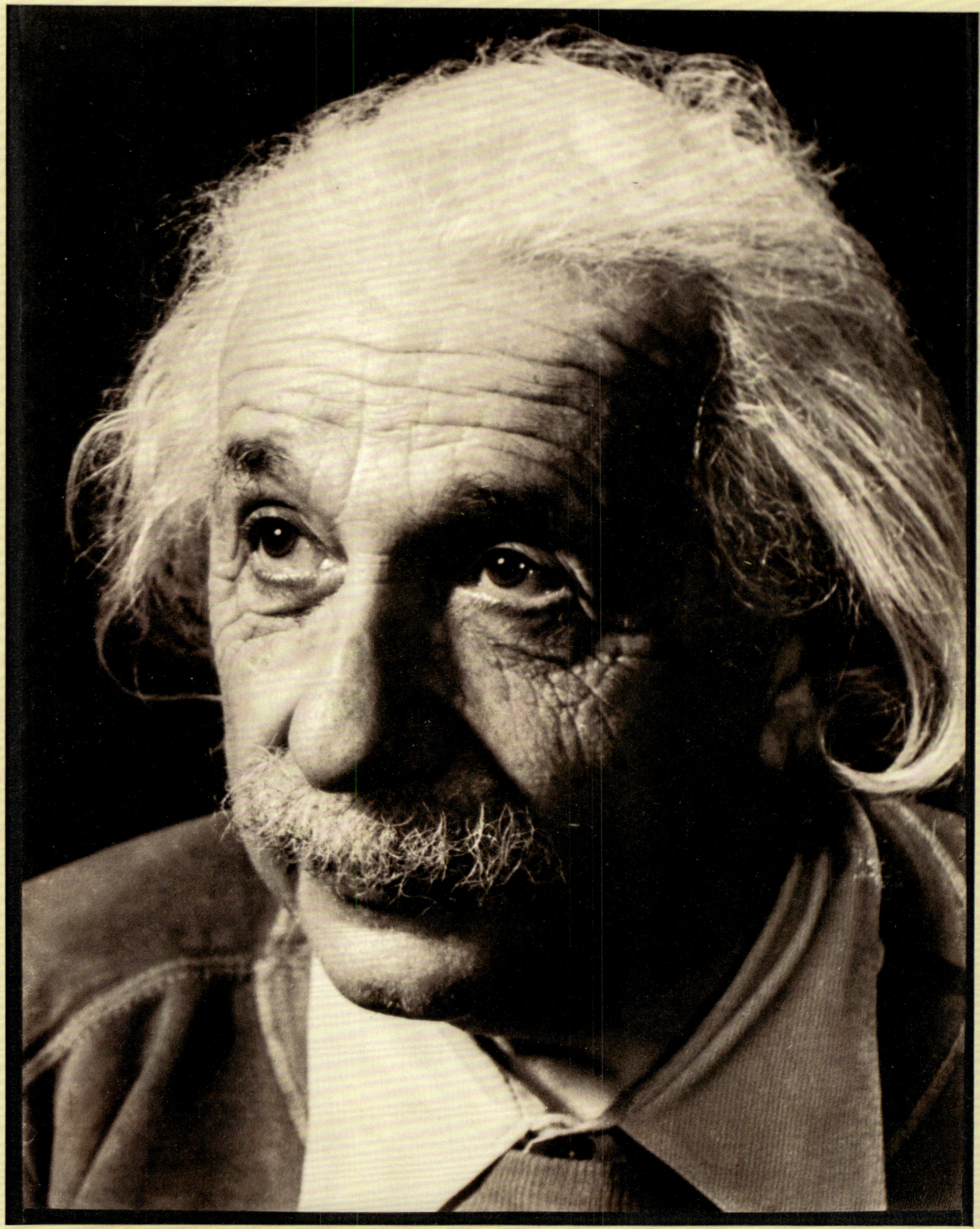

Herrn Sternberger A. Einstein. 50.

marcel sternberger

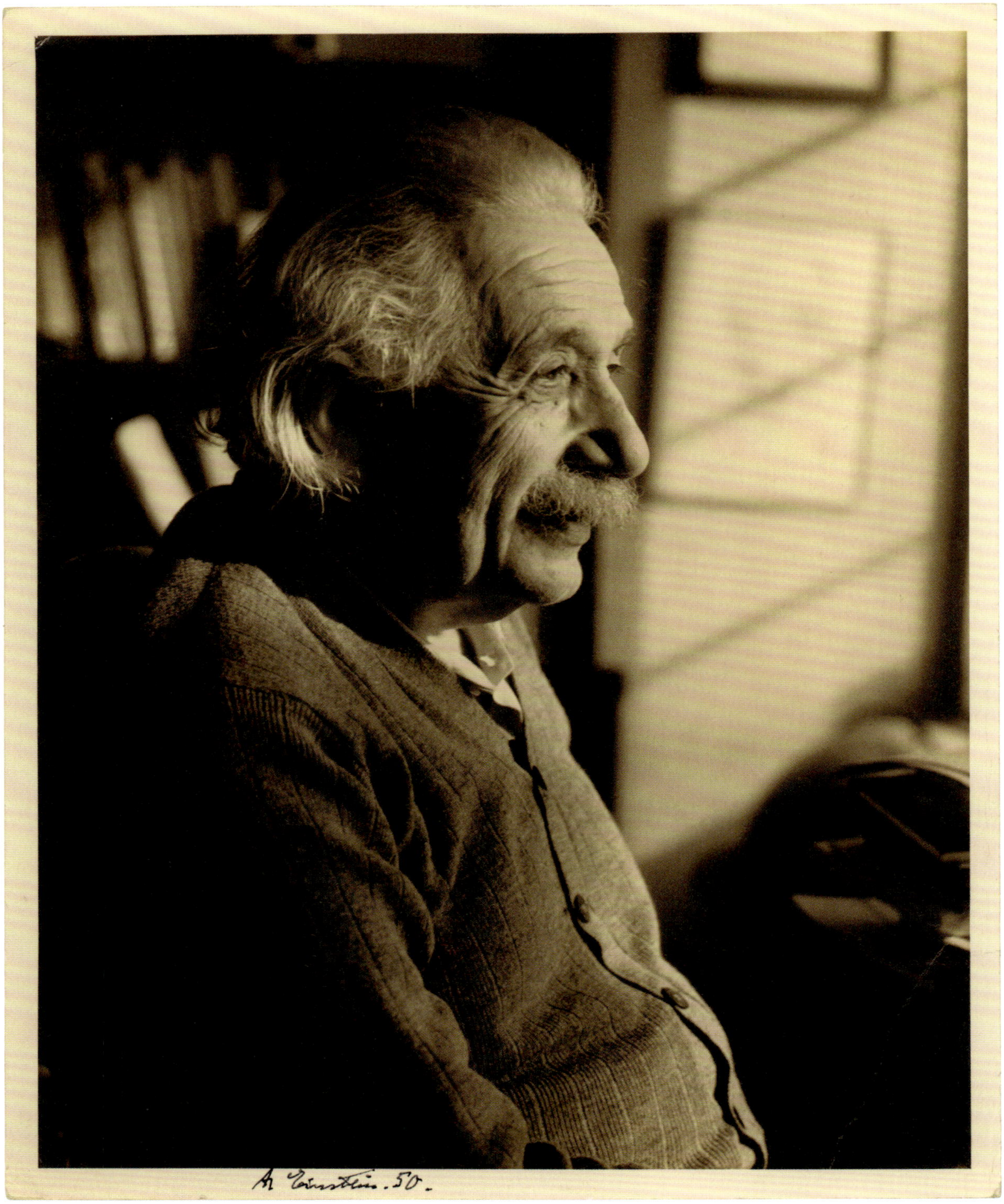
A. Einstein. 50.

## *SERENE DESPITE POLITICAL ATTACKS*

Sitting comfortably in his Princeton study, Einstein looks relaxed and is smiling in this beautiful profile portrait. Einstein signed it in 1950, a time that should have been peaceful for a man who had accomplished so much in life and was celebrated by his scientific peers and the public for his genius.

The year 1950, however, saw the beginnings of McCarthyism in the United States. Although Einstein often said he never liked to get involved in politics, because of his fame any political remarks he made were amplified by the press and quickly became consequential.

As a pacifist in favor of nuclear disarmament, a consistent supporter of civil rights, and at times a supporter of socialism, Einstein was considered a radical and regarded with suspicion by a number of people in the US government in the 1950s. During the height of McCarthyism, Einstein encouraged intellectuals not to cooperate with the House Un-American Activities Committee, arguing the committee violated individual liberties including the freedom of thought and speech. He thus became an instant lightning rod for controversy, subjected to scathing articles in the press that questioned his loyalty. One major newspaper even suggested that Einstein was being used as an instrument of propaganda by enemies of the United States. The US government took notice, and although he was never called before an investigatory committee, the FBI kept a file on him that grew to 1,427 pages.

UNKNOWN PHOTOGRAPHER,
*1950. Signed and dated by Einstein.*

## *IN HIS GARDEN WITH A FRIEND*

Here Einstein is pictured, at age seventy-two, with a member of his inner circle, Alice Loewy Kahler. Kahler was married to the Czech Jewish philosopher and literary scholar Erich Kahler. The Kahler home at One Evelyn Place in Princeton was a center for intellectual conversation and particularly welcomed Jewish scholars who had fled Europe during the war. Members of the "Kahler Circle," as it came to be known, included other luminaries such as Thomas Mann, Kurt Gödel, Ben Shahn, and Hermann Broch.

This photo, taken by fellow Jewish refugee Trude Fleischmann, is signed by Fleischmann and inscribed by Einstein to Kahlers' daughter, Hanna.

PHOTOGRAPH BY TRUDE FLEISCHMANN,
*Princeton, NJ, 1951. Signed by the photographer and signed, dated, and inscribed by Einstein: "For Hannah."*

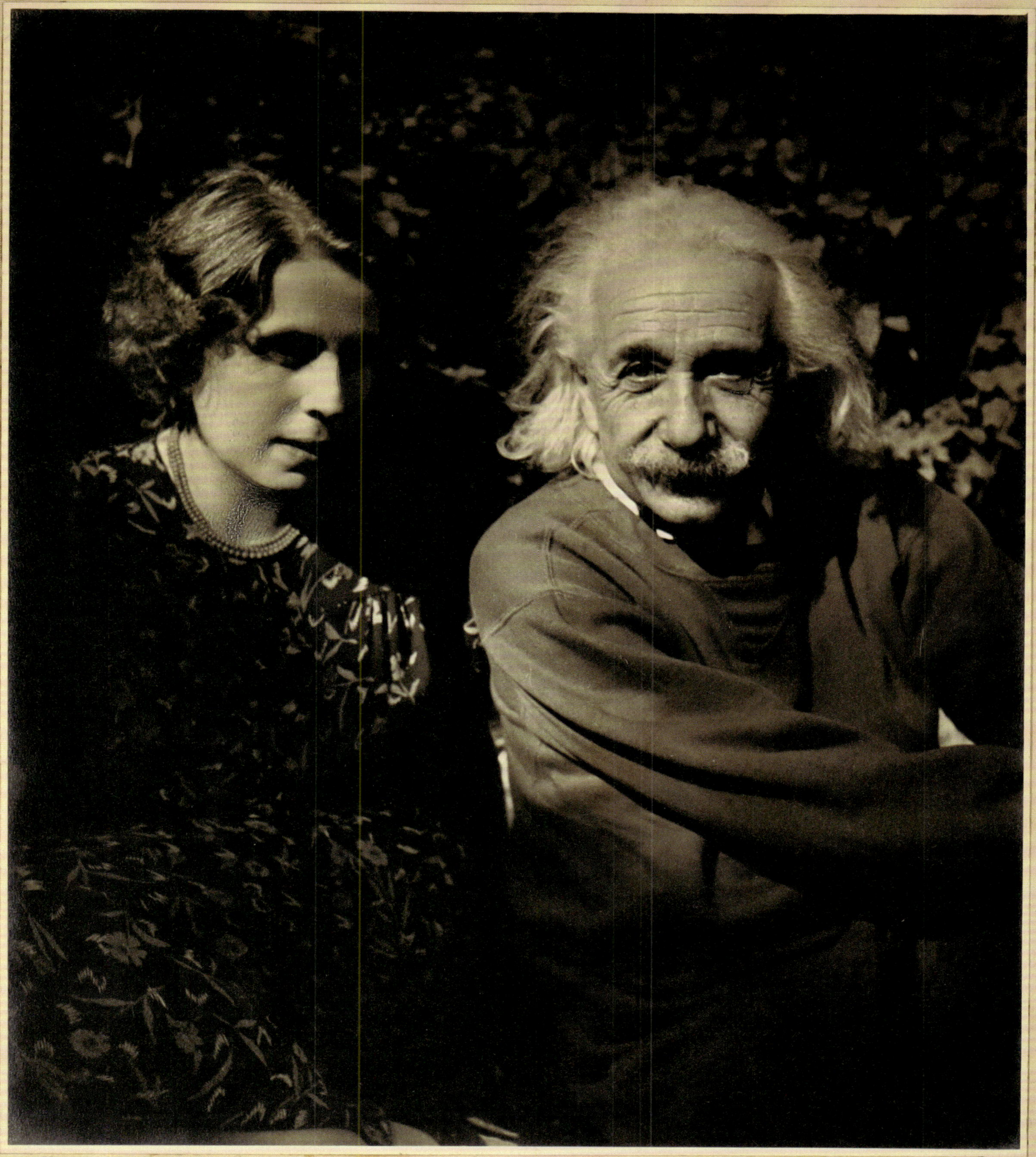

Trude Fleischmann
New York

Für Hamerl
A. Einstein. 51.

### *WALL OF HEALING*

---

This striking portrait shows Einstein relaxed and smiling at his home in Princeton during a ceremony to honor the groundbreaking of the Hebrew University-Hadassah Medical Centre in Jerusalem.

During the ceremony, Hadassah (the Women's Zionist Organization of America) and the American Friends of the Hebrew University gave Einstein a scroll documenting the inscription of his name planned for the "Wall of Healing" at the new medical center. According to newspaper reports of the occasion, Einstein noted that the center "is of great significance to Israel itself, its relation to the neighboring countries and the status of Jews everywhere."

Einstein was an advocate of both the State of Israel and Jewish-Arab cooperation. When Israel's first president, Chaim Weizmann, died in 1952, Einstein was offered the presidency, which he refused.

Einstein signed and dated this photograph in the lower margin.

PHOTOGRAPH BY NATHANIEL FEIN,
*Princeton, NJ, 1952. Signed and dated by Einstein.*

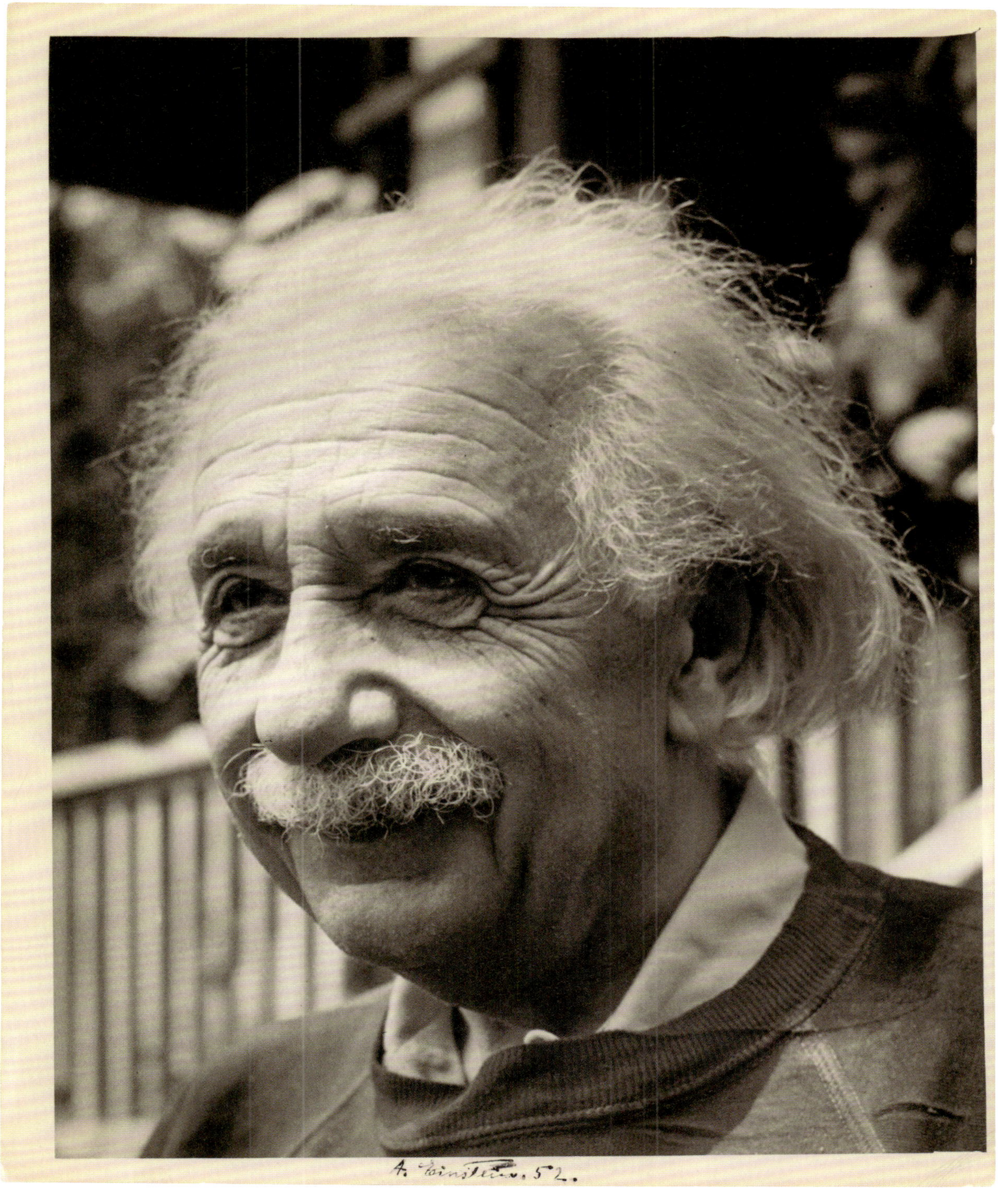
A. Einstein. 52.

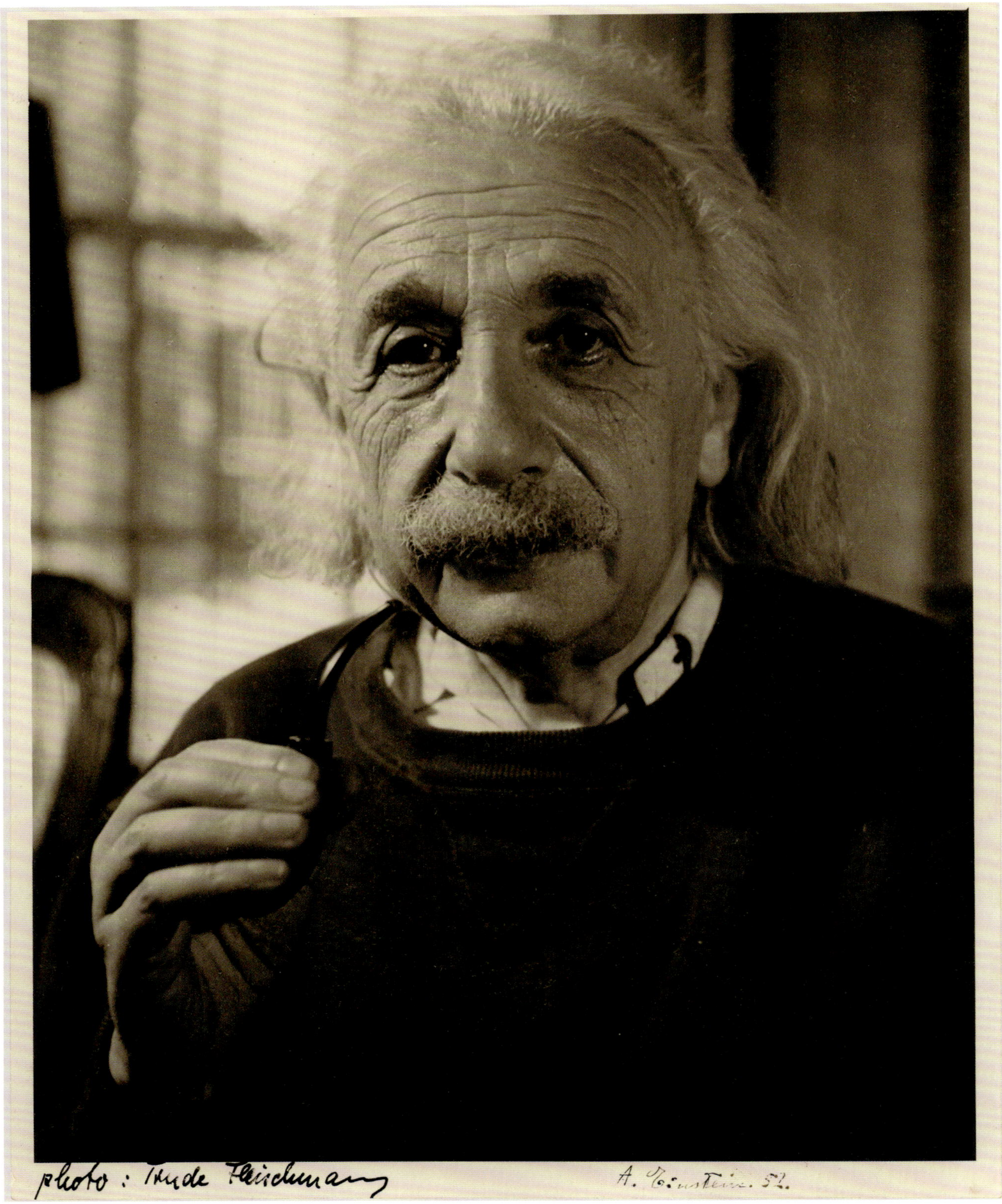
photo: Trude Fleischmann
A. Einstein. 52.

### *INTERVENING YEARS*

This photograph by Trude Fleischmann is reminiscent of her 1946 photo of Einstein. While Einstein's facial expression, attire, and pose, holding pipe in hand, are quite similar in both images, the intervening years had taken a toll.

In 1948, doctors discovered that Einstein had a large aortic aneurysm. There was no cure for the condition. His doctors recommended that he adopt a healthier lifestyle—a better diet and no smoking—to reduce the chances of a fatal rupture of the artery.

Einstein signed and dated the white border of this print in ink. An unidentified hand has added to the left of his signature: "photo: Trude Fleischmann." The photographer's stamps bearing Fleischmann's name and address are on the verso.

PHOTOGRAPH BY TRUDE FLEISCHMANN,
*Princeton, NJ, 1952. Signed and dated by Einstein.*

## *RED, TREE, DISTANCE, ATOM*

In his 1930 letter to Dr. Ralph Lyndal Worrall (discussed on page 82), Einstein explained that all physics must start from the "hypothesis of a reality that is independent of perception and thought."

Two decades later, Einstein returns to the issue of perception and reality. He gives a more detailed explanation of his approach to science in these two letters to Colombian mathematician, philosopher, and politician Mario Laserna Pinzón.

In the first letter, dated September 22, 1953, Einstein writes:

"Concepts, as far as they have any basis, are—judged logically—free inventions of the mind (together with propositions connecting them). But those concepts and propositions receive their value and justification exclusively through their only intuitively given connection with perceptions (Erlebnissen). There is no logical way to deduce concepts and propositions from our crude experiences ('induction'). This is equally true for concepts like 'red,' 'tree,' as for concepts like 'distance,' 'atom,' etc. The difference lies in the fact that scientific concepts and propositions are mostly brought into connection with sense-perceptions in a more indirect and complicated way."

In his second letter, dated January 8, 1955, Einstein is more explicit:

"It is basic for all physics that one assumes a real world existing independently from any act of perception. But this we do not know. We take it only as a programme in our scientific endeavours. This programme is, of course, pre-scientific and our ordinary language is already based on it.

"The concepts body-object and shape are not given to us directly by our sense-impressions but are a result of a mental construct. That this is not so easy to see is only produced by the fact that those steps made by every one of us in early childhood seem to us logically necessary. But this is not so."

We find it remarkable that Einstein—famous for explaining how the universe behaves—is saying in these letters that we can only *assume*, but not *know*, that a real world exists outside of our perceptions of it. The second letter was written only three months before Einstein's death. It expresses what is likely his final philosophical view underlying all of his work.

CORRESPONDENCE, *Einstein to Mario Laserna Pinzón, September 22, 1953.*

THE INSTITUTE FOR ADVANCED STUDY

PRINCETON, NEW JERSEY

September 22,1953

SCHOOL OF MATHEMATICS

Dr.Mario Laserna
Universidad de los Andes
Calle 18-A Carrera 1-E
Apartado Aereo 4976
Bogota,Colombia.

Dear Mr.Laserna:

I gather from your dialogue that a considerable difference of opinion exists between us with regard to this problem. Because I am not of the opinion that there exists an essential difference between concepts and methods in the fields of "common sense" and science.

Every linguistic utterance is wholly confined to the conceptual sphere. Concepts, as far as they have any basis, are -judged logically - free inventions of the mind (together with propositions connecting them). But those concepts and propositions receive their value and justification exclusively through their only intuitively given connection with perceptions (Erlebnissen). There is no logical way to deduce concepts and propositions from our crude experiences ("induction"). This is equally true for concepts like "red", "tree",as for concepts like "distance", "atom", etc. The difference lies in the fact that scientific concepts and propositions are mostly brought into connection with sense-perceptions in a more indirect and complicated way. Also the use of numbers does not involve a difference in essence between scientific and common sense methods.

Apart from these differences of opinion I must confess that I do not want to appear in this field as a responsible partner. I have not studied epistomology thouroughly enough and I am not sufficiently aquainted with the tremendously extented literature in that field. It is enough if you mention that we had several discussions about these questions.

With kind regards,

A. Einstein.

Albert Einstein.

$$G_{\mu\nu} = \frac{8\pi G}{c^4} T_{\mu\nu}$$

## *THE ACQUISITION OF KNOWLEDGE*

The historian of philosophy Max Fishler wrote to Einstein in 1953 asking for his thoughts about the philosophy of Immanuel Kant. Fishler was planning to write a book on Kant's views as they pertained to the new knowledge of relativity. Einstein's thoughts, of course, would be essential.

Einstein had studied Kant's philosophy in depth, and it had profoundly influenced his early thinking. In this revealing letter shown on the following pages, Einstein responds to Fishler by rejecting one of Kant's central tenets regarding the acquisition of knowledge.

Einstein begins the letter with a discussion of Kant's work, emphasizing that Kant accepted as valid "a priori" knowledge—knowledge acquired independently of experience. Einstein rejects Kant's premise, arguing that knowledge based purely on reason is inherently uncertain and liable to be overturned. He cites the development of non-Euclidean geometry and his own work as examples disproving Kant's premise.

Translated from the original German, the letter reads, in part:

"We may say then—as I see it, with justice—that all concepts, not only the mathematical ones, are a priori insofar as they are not logically deducible from naked experience. But this holds for all concepts, for the concepts of the empirical sciences not less than for the concepts of pure mathematics.

"The general theory of relativity has in my opinion convincingly shown that the spatial character belongs to the objects of the physical world as a mere characteristic of it (i.e. four-dimensionality of space-time).

"The contrast to Kant can be illustrated briefly through the short sentence: the inner space of a box is 'real' in the same sense as the box itself."

Einstein dwelled deeply on the philosophical implications and requirements of his theories, making major contributions to the philosophy of science. One scholar noted that "Einstein's influence on twentieth-century philosophy of science is comparable to his influence on twentieth-century physics."

*Pages 186–187:* CORRESPONDENCE,
*Einstein to Max Fishler, December 17, 1953.*

den 17.Dezember 1953

Herrn Max Fishler
336 No.Sierra Bonita Ave.
Los Angeles 36,Cal.

Sehr geehrter Herr Fishler:

Ich danke Ihnen freundlich für die Uebersendung der Gramophonplatte, die in gutem Zustand eingetroffen ist. Ich werde sie mir anhören sobald ich Gelegenheit habe.

Die Frage wegen Zeit und Raum im Zusammenhang mit Kant's Philosophie ist insofern nicht ganz leicht zu beantworten als die Auffassung Kants bezüglich Zeit und Raum von verschiedenen Leuten verschieden interpretiert wird. Mir aber scheint das Wesentliche darin zu liegen, dass Kant folgende Auffassung vertritt:

Das räumliche Denken ist nicht in dem selben Sinne an die sinnlichen Erfahrung gebunden, wie etwa das Denken bezüglich körperlicher Gegenstände. Die räumlichen Begriffe sind für ihn a priori, also vor aller Erfahrung gegeben und gewissermassen angeborene Werkzeuge der Wahrnehmung und des Denkens. (Anschauung a priori). Auf diesen Sachverhalt führt er die "Unbezweifelbarkeit" geometrischer Sätze zurück.

Heute aber zweifelt wohl kaum jemand daran, dass die Unbezweifelbarkeit nur in dem Sinne besteht, dass bezw.insofern es sich um logische Folgerungen aus vorgegebenen Axiomen handelt, welche selbst from logischen Standpunkte willkürlich gesetzt sind.

Natürlich ist diese Einsicht heute trivial geworden, seitdem die Mathematiker andere Geometrien aufgestellt haben, die von der euklidischen Geometrie abweichen und logisch ebenso folgerichtig sind wie die euklidische Geometrie.

Wir wissen ferner, dass die Setzung der euklidischen Axiome durch unsere Erfahrungen an festen Körpern angeregt worden sind, sodass also eine <u>psychologische</u> Abhängigkeit von empirischen Gegebenheiten kaum zu bezweifeln ist.

Nun kann man allerdings - nach meiner Ansicht mit Recht - sagen, dass überhaupt alle Begriffe, also nicht nur die mathematischen, insofern a priorisch sind, als sie nicht auf logische Weise aus den

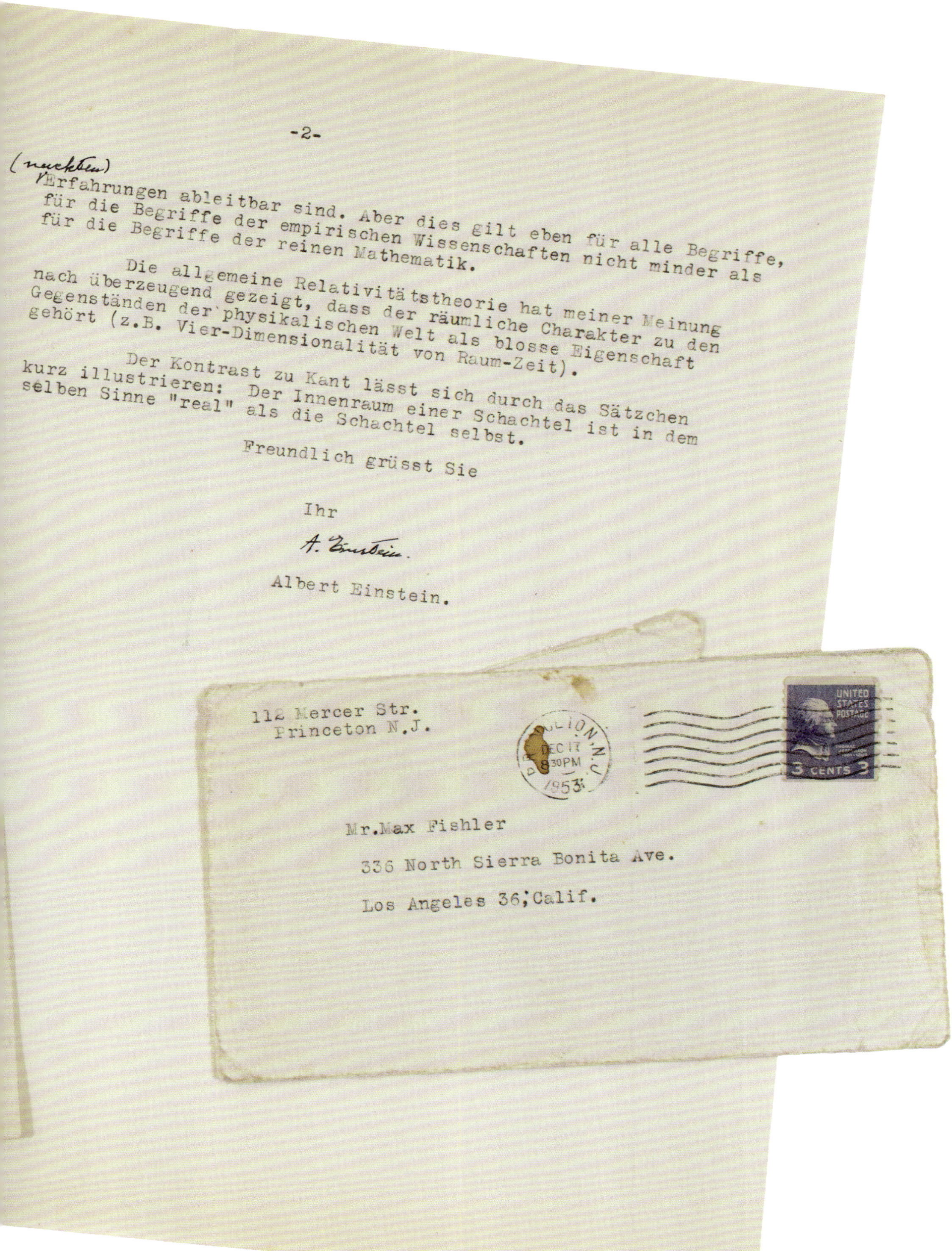

-2-

(nackten) Erfahrungen ableitbar sind. Aber dies gilt eben für alle Begriffe, für die Begriffe der empirischen Wissenschaften nicht minder als für die Begriffe der reinen Mathematik.

Die allgemeine Relativitätstheorie hat meiner Meinung nach überzeugend gezeigt, dass der räumliche Charakter zu den Gegenständen der physikalischen Welt als blosse Eigenschaft gehört (z.B. Vier-Dimensionalität von Raum-Zeit).

Der Kontrast zu Kant lässt sich durch das Sätzchen kurz illustrieren: Der Innenraum einer Schachtel ist in dem selben Sinne "real" als die Schachtel selbst.

Freundlich grüsst Sie

Ihr

A. Einstein.

Albert Einstein.

112 Mercer Str.
Princeton N.J.

DEC 17 8 30 PM 1953

UNITED STATES POSTAGE 3 CENTS 3

Mr.Max Fishler

336 North Sierra Bonita Ave.

Los Angeles 36, Calif.

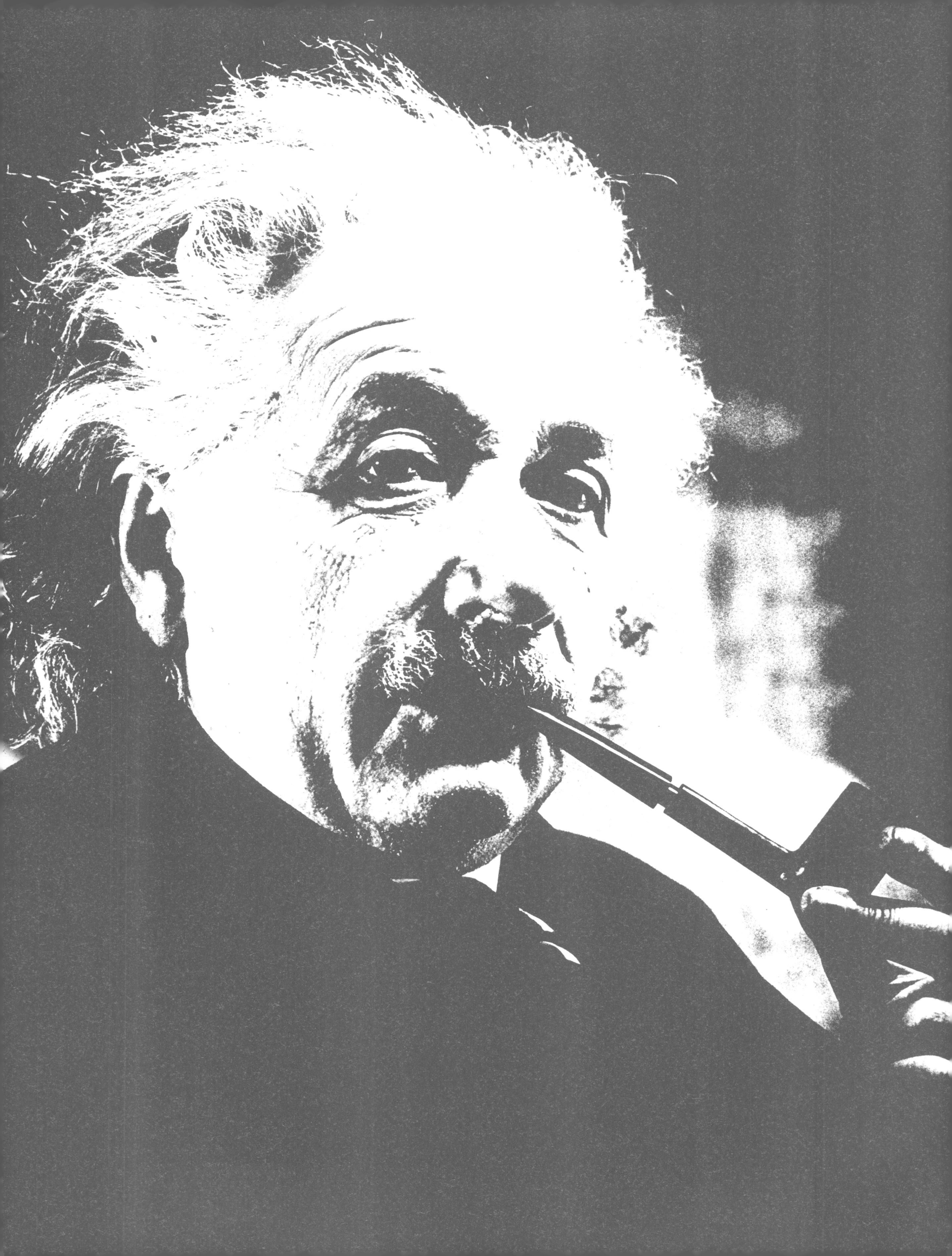

*It is basic for all physics that one assumes a real world existing independently from any act of perception. But this we do not know."*

—ALBERT EINSTEIN, *1955*

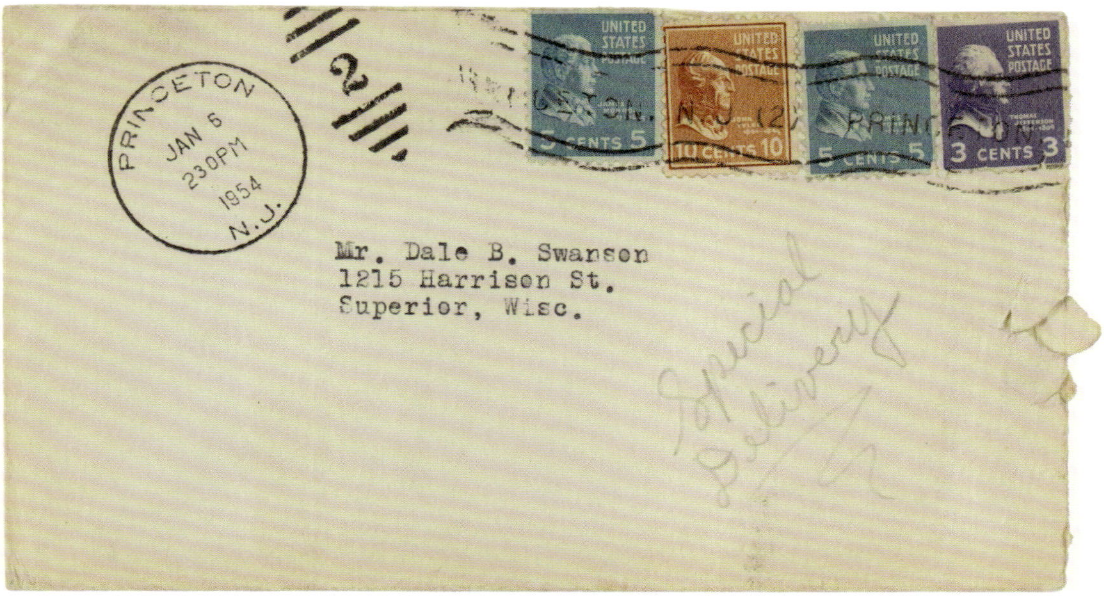


## *MASS, ENERGY, AND LIGHT*

Although by 1954 he was one of the most recognized figures in the world, Einstein would still take time out of his busy schedule to answer questions from fans, students, and amateur scientists who were eager to discuss his scientific theories. He did not ignore or belittle his public but sincerely tried to help them understand his work, corresponding—as a good teacher would—in clear, precise language, free of complicated equations and obscure terminology.

In this letter, written near the end of his life, Einstein responds to a question concerning one of the most fundamental aspects of his life's work—the relationship between mass, energy, and light. The letter reads:

"Dear Sir:
One should speak of mass only in the sense of something characteristic for the body and independent of its motion [rest mass]. Now it is true that the energy of a finite mass becomes infinite [in motion], provided the mass is finite. In the case of the photon the mass has to be assumed zero or infinitely small in such a way that in spite of its having the light velocity its energy is finite.

Sincerely yours,
[Signed] A. Einstein.
Albert Einstein."

CORRESPONDENCE,
*Einstein to Dale B. Swanson, 1954.*

THE INSTITUTE FOR ADVANCED STUDY
PRINCETON, NEW JERSEY

Januar 5,1954

SCHOOL OF MATHEMATICS

Mr.Dale B.Swanson
1215 Harrison Str.
Superior,Wisc.

Dear Sir:

One should speak of mass only in the sense of something characteristic for the body and independent of its motion. Now it is true that the energy of a finite mass becomes infinite, provided the mass is finite. In the case of the photon the mass has to be assumed zero or infinitely small in such a way that inspite of its having the light velocity its energy is finite.

Sincerely yours,

A. Einstein.

Albert Einstein.

## *THE UNGUARDED MOMENT*

In this photograph taken one year before his death, Einstein is seated in his Princeton home surrounded by books, holding his pipe while gazing slightly away from the camera. He signed and dated the print.

In his book *The Unguarded Moment*, the photographer Frederick Plaut describes taking this evocative photo of the elderly Einstein:

"There must be a moment in every professional photographer's life when he is so in awe of his subject that he can scarcely focus his camera. That moment for me was when I met Albert Einstein at his home in Princeton. Certainly the great man was not formidable; he greeted my wife and me graciously, and proceeded to chat with her while I went to work. I remember that she asked him about his music and when he told her that he no longer played his violin she murmured, 'That's too bad.' He smiled, 'Ah, no. It would have been too bad if I went on.' In the final moments of our visit, Einstein looked at me very seriously. 'I hope,' he said, 'you can sell these pictures for a good price.' Astounded, I blurted out: 'Oh, no, Sir. I have nothing to sell. I just wanted to photograph you.' His face clouded. 'Not sell them? If I had known that I never would have let you take them.' After we left, I realized the significance of a delightful remark attributed to Mrs. Einstein. Someone once asked Mrs. Einstein whether she understood Professor Einstein's theory of relativity. She answered without hesitation, 'No, but I understand Professor Einstein.'"

PHOTOGRAPH BY FREDERICK PLAUT,
*Princeton, NJ, 1954. Signed and dated by Einstein.*

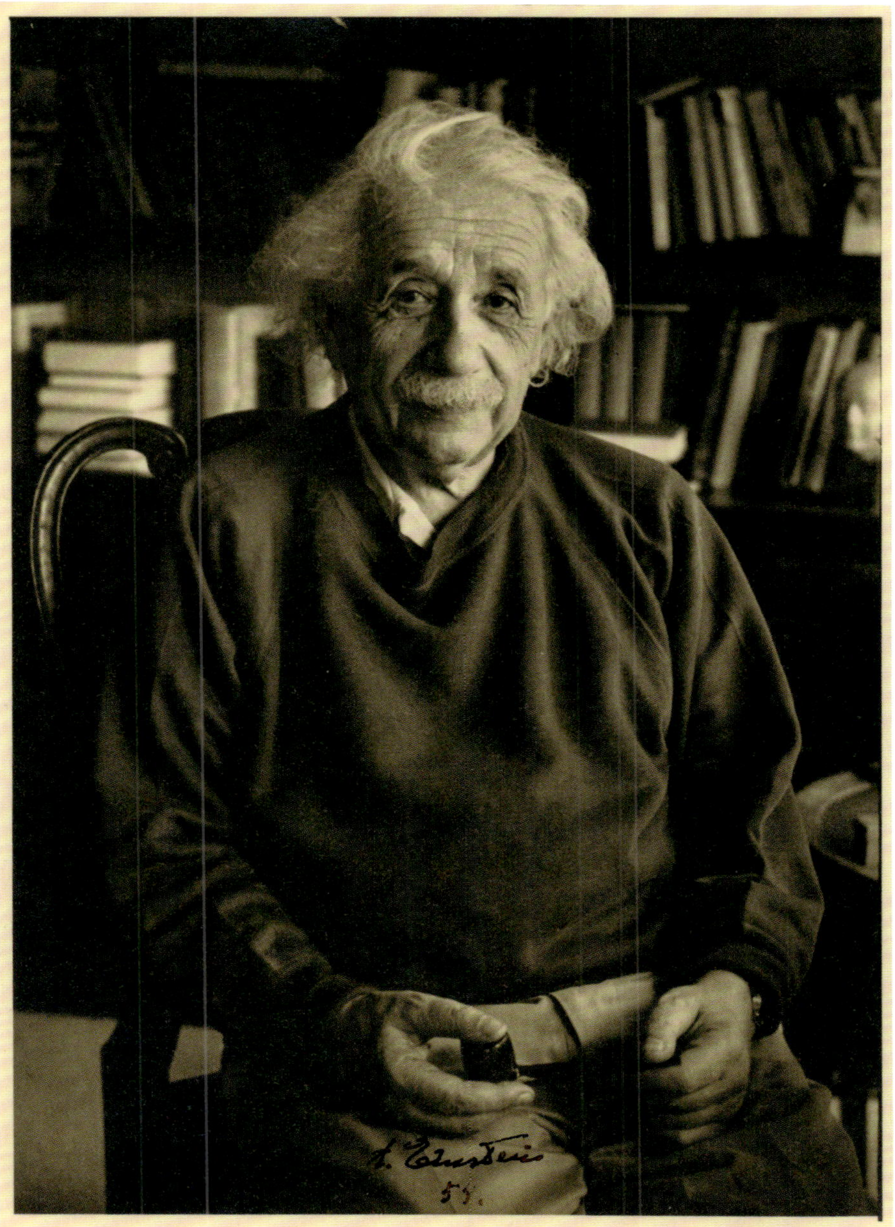

## *THE LAST SIGNED PHOTOGRAPH*

We believe this to be Einstein's last signed photograph.

The recipient was Andrew J. Robell, a student at Princeton in 1955. In an affidavit that accompanies the photograph, he explained the circumstances surrounding Einstein's signature:

"I learned that Albert Einstein lived at 112 Mercer Street in Princeton, a short distance from my freshman dormitory. Early in 1955 (January or February, as I recall) I walked to his house carrying the accompanying photograph of Dr. Einstein and asked that he sign it."

Einstein signed and dated the photograph in black ink.

We have not been able to identify the photographer definitively, but it bears the characteristics of Orren Jack Turner II, whose father had photographed Einstein in Princeton in 1921 during Einstein's first visit to the United States.

For decades, Einstein had suffered from abdominal pain. In December 1948, he underwent exploratory surgery that revealed a large aortic aneurysm. The surgeon, Dr. Rudolph Nissen, wrapped the bulging blood vessel with cellophane to stimulate fibrosis and reinforce the weakened vessel. (Resecting the defective part and replacing it with a synthetic graft was not yet possible.) The operation added over six years, mostly pain-free, to Einstein's life.

On April 12, 1955, Einstein experienced severe abdominal pain that intensified the following day. At first, he refused hospitalization but soon agreed to be admitted to Princeton Hospital to avoid being a burden at home. Einstein's aneurysm was leaking and in the early stage of rupturing. He refused surgery, knowing that he would soon die without it, saying: "I want to go when I want. It is tasteless to prolong life artificially. I have done my share, it is time to go. I will do it elegantly."

Einstein continued his life's work during the five days he was hospitalized. Early in the morning of April 18, 1955, the aneurysm ruptured. Einstein uttered his last words in German—which his English-speaking night nurse did not understand—and his extraordinary life ended.

UNKNOWN PHOTOGRAPHER
*(possibly Orren Jack Turner II).*
*Signed and dated by Einstein in 1955.*

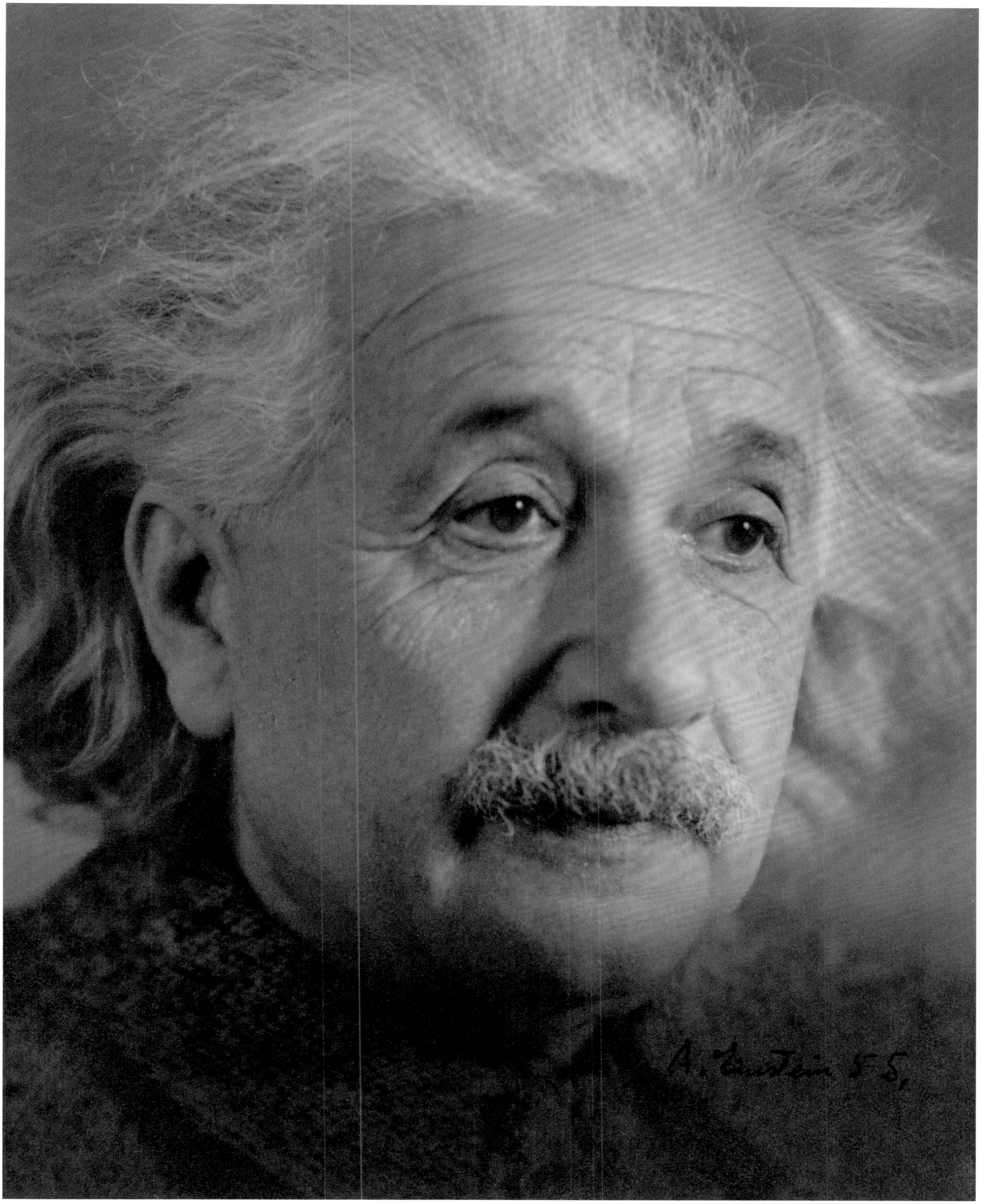
A. Einstein 55,

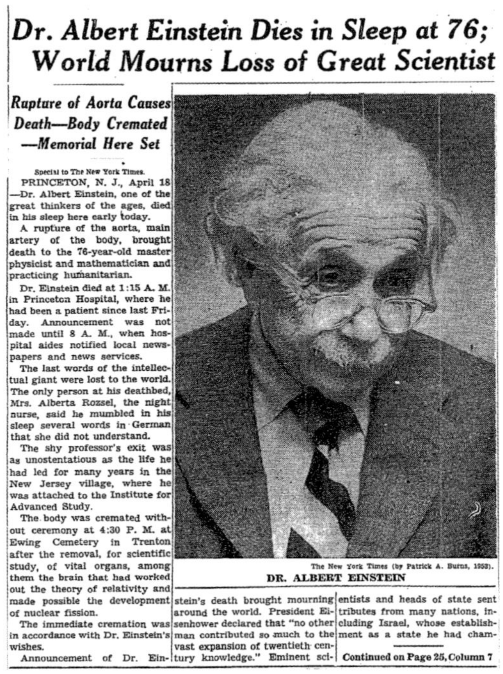

**Dr. Albert Einstein Dies in Sleep at 76; World Mourns Loss of Great Scientist**

***Rupture of Aorta Causes Death—Body Cremated —Memorial Here Set***

Special to The New York Times.

PRINCETON, N. J., April 18—Dr. Albert Einstein, one of the great thinkers of the ages, died in his sleep here early today.

A rupture of the aorta, main artery of the body, brought death to the 76-year-old master physicist and mathematician and practicing humanitarian.

Dr. Einstein died at 1:15 A. M. in Princeton Hospital, where he had been a patient since last Friday. Announcement was not made until 8 A. M., when hospital aides notified local newspapers and news services.

The last words of the intellectual giant were lost to the world. The only person at his deathbed, Mrs. Alberta Rozsel, the night nurse, said he mumbled in his sleep several words in German that she did not understand.

The shy professor's exit was as unostentatious as the life he had led for many years in the New Jersey village, where he was attached to the Institute for Advanced Study.

The body was cremated without ceremony at 4:30 P. M. at Ewing Cemetery in Trenton after the removal, for scientific study, of vital organs, among them the brain that had worked out the theory of relativity and made possible the development of nuclear fission.

The immediate cremation was in accordance with Dr. Einstein's wishes.

Announcement of Dr. Einstein's death brought mourning around the world. President Eisenhower declared that "no other man contributed so much to the vast expansion of twentieth century knowledge." Eminent scientists and heads of state sent tributes from many nations, including Israel, whose establishment as a state he had cham-

Continued on Page 25, Column 7

The New York Times (by Patrick A. Burns, 1953).

DR. ALBERT EINSTEIN

## *DR. EINSTEIN DIES IN SLEEP AT 76*

The day after he died, the *New York Times* published Einstein's obituary at the top of the front page. His picture appeared prominently over two columns, under the headline, "Dr. Einstein Dies in Sleep at 76."

*Times* photographer Patrick A. Burns had taken the photograph two years earlier, in 1953. In German, Einstein inscribed and signed the print shown here: "For Mrs. Nina Rubenstein / A. Einstein. 54." (Rubenstein had fled Nazi Germany in 1933 and later settled in New York, becoming a UN interpreter.)

Burns's photograph perfectly captured the aged Einstein, hunched slightly forward, appearing vulnerable and so very human.

The *Times* obituary noted, "As the years passed, the figure of Einstein the man became more and more remote, while that of Einstein the legend came ever nearer to the masses of mankind. . . . 'Saintly,' 'noble' and 'lovable' were the words used to describe him by those who knew him even casually. He radiated humor, warmth and kindliness. He loved jokes and laughed easily.

"Princeton residents would see him walk in their midst, a familiar figure, yet a stranger, a close neighbor, yet at the same time a visitor from another world. And as he grew older his otherworldiness became more pronounced, yet his human warmth did not diminish."

PHOTOGRAPH BY PATRICK A. BURNS,
*Princeton, NJ, 1953. Signed, dated, and inscribed by Einstein in 1954.*

*I want to go when I want.*
*It is tasteless to*
*prolong life artificially.*
*I have done my share,*
*it is time to go.*
*I will do it elegantly."*

—ALBERT EINSTEIN, *1955*

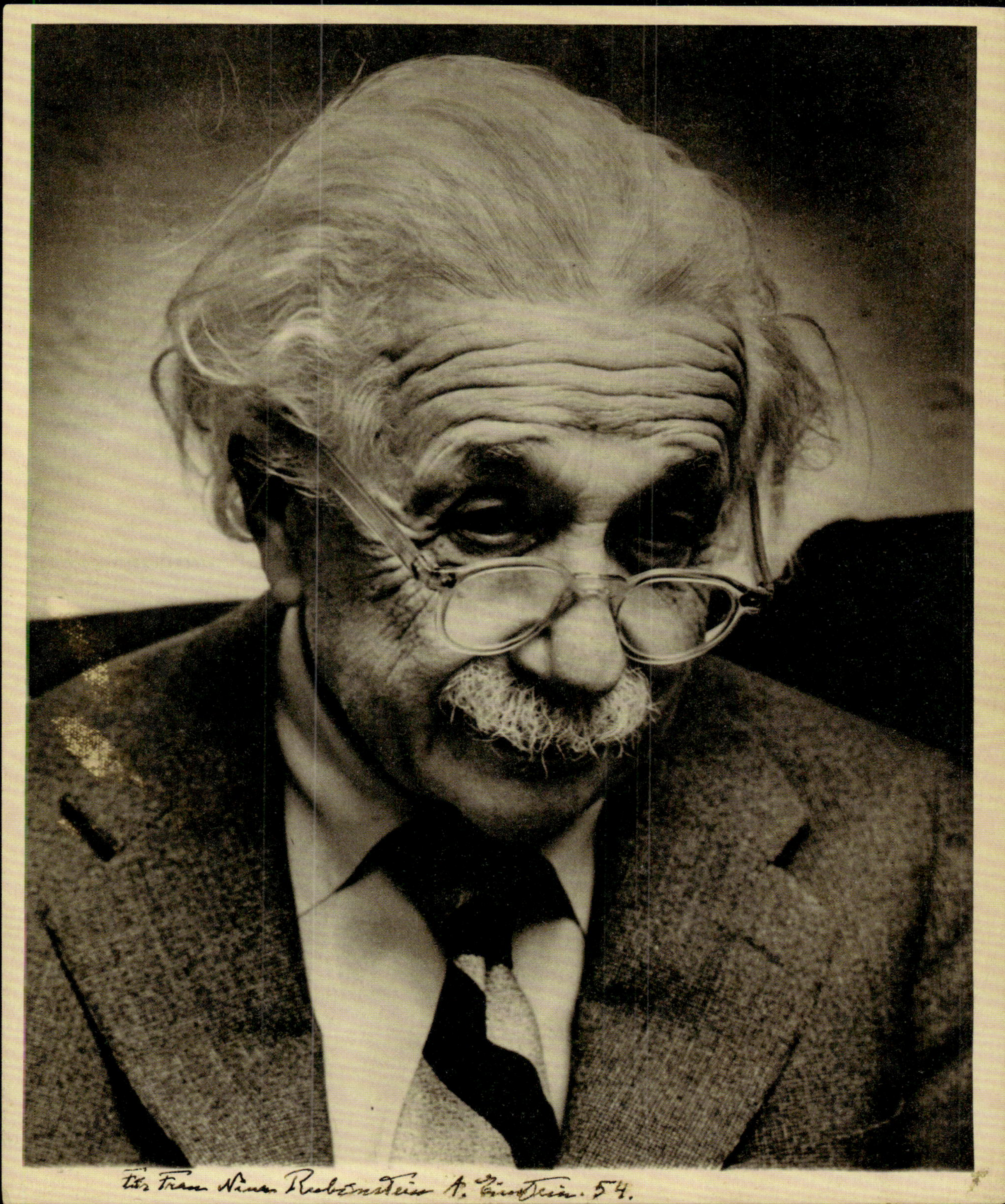
für Frau Nina Rubinstein A. Einstein. 54.

# *EPILOGUE*

In Midtown Manhattan, there is a giant mural of Einstein by the Brazilian artist Eduardo Kobra. If you walk by, you can't miss it. Here is Einstein, several stories high, on the side of a building, cheerfully riding a bicycle displaying a sign reading ☮ = ♥². It's stunning. Even jaded New Yorkers sometimes stop to take pictures in front of it.

Kobra has created murals of other figures around the city: Elvis, Michael Jackson, David Bowie, Jim Morrison, Jimi Hendrix, Janis Joplin, and more. These all make sense—they were literally rock stars. But a scientist dressed in a tie and plain sweater who has been dead for nearly seventy years? That's odd. But the strangest thing is, perhaps, that it doesn't seem so strange, for Einstein's image is everywhere—in college dorm rooms, in advertisements, in fine art galleries, and on the streets of New York.

What is it about Einstein? What has made his image endure all these years, even for many who know little of his scientific achievements?

In helping to curate the Berger collection, I often found myself staring at an Einstein photo wondering what makes his image so powerful. Whether he is trying to solve a difficult scientific problem, full of despair contemplating the fate of the world, or lightheartedly playing with children, Einstein seems to be communicating his emotions directly to us. Somehow, in looking at these photographs we feel we know him, that if he walked into the room right now we could talk to him and understand each other. This is extraordinary, considering we are contemplating someone who explored realms of thought inaccessible to nearly all of us.

Perhaps that is part of his appeal as well—he seems approachable and often empathetic, yet when we look at him we are also filled with wonder and inspired by his genius.

I feel privileged to have worked on the Berger collection for many years. It was joyful to be surrounded by these images and to study Einstein's thoughts and discoveries through the original documents and letters. We hope in these pages, we were able to communicate and share this joy with you.

—MICHAEL DiRUGGIERO

## *WORKS CITED*


PAGE 9
Viereck, George Sylvester. "What Life Means to Einstein: An Interview by George Sylvester Viereck." *Saturday Evening Post*, October 26, 1929.

PAGE 17
Isaacson, Walter. *Einstein: His Life and Universe*. New York: Simon & Schuster, 2007, 64.

PAGE 21
Isaacson, Walter. *Einstein: His Life and Universe*. New York: Simon & Schuster, 2007, 26.

PAGE 28
Brian, Denis. *Einstein: A Life*. New York: John Wiley & Sons, Inc., 1996, 82.

PAGE 29
Brian, Denis. *Einstein: A Life*. New York: John Wiley & Sons, Inc., 1996, 82.

PAGE 30
Pais, Abraham. *Subtle Is the Lord: The Science and Life of Albert Einstein*. Oxford: Oxford University Press, 2005, 250.

PAGE 32
American Museum of Natural History. "Course Author: Albert Einstein." Seminars on Science. https://www.amnh.org/learn-teach/seminars-on-science/about/faculty/albert-einstein.

Robinson, Andrew. *Einstein: A Hundred Years of Relativity*. Princeton: Princeton University Press, 2015, 196.

Isaacs, Garry P. *Standing Up to the Coup: A Fight to Preserve Freedom and Human Health*. Self-published, 2021, 48.

INSERT BEFORE PAGE 33
This equation expresses the relationship between mass and energy.

PAGE 34
Christie's New York. *Einstein: Letters to a Friend, Part II*. Online auction, 2017, lot 8.

Isaacson, Walter. *Einstein: His Life and Universe*. New York: Simon & Schuster, 2007, 26.

Kaku, Michio. *Parallel Worlds: A Journey Through Creation, Higher Dimensions, and the Future of the Cosmos*. New York: Knopf Doubleday Publishing Group, 2006, 39.

PAGES 42–43
Einstein, Albert, and N. Rosen. "On Gravitational Waves." *Journal of the Franklin Institute* 233, no. 1 (January 1937).

PAGE 49
Einstein, Albert. *Relativity: The Special and the General Theory*. Translated by Robert W. Lawson. New York: Henry Holt and Company, 1920, 65.

PAGE 50
Calaprice, Alice. *An Einstein Encyclopedia*. Princeton: Princeton University Press, 2015, 86.

Calaprice, Alice. *The Einstein Almanac*. Baltimore: The Johns Hopkins University Press, 2005, 50.

PAGE 54
Einstein, Albert. *Relativity: The Special and the General Theory*. Translated by Robert W. Lawson. New York: Henry Holt and Company, 1920, 40–41.

PAGE 56
National Portrait Gallery. "Walter Benington: Biography." Accessed December 24, 2021. https://www.npg.org.uk/whatson/display/2007/walter-benington-pictorial-portaits/biography.

Roazen, Paul. "Brunswick, Ruth Jane Mack (1897–1946), Psychoanalyst." *American National Biography Online* (February 2000).

PAGE 58
Princeton University Press. "The Collected Papers of Albert Einstein." https://einsteinpapers.press.princeton.edu/.

PAGE 60
Pais, Abraham. *Subtle Is the Lord: The Science and Life of Albert Einstein*. Oxford: Oxford University Press, 1982, 443.

INSERT BEFORE PAGE 61
Pais, Abraham. *Subtle Is the Lord: The Science and Life of Albert Einstein*. Oxford: Oxford University Press, 1982, 443.

PAGE 62
Isaacson, Walter. *Einstein: His Life and Universe*. New York: Simon & Schuster, 2007, 326.

PAGE 65
Einstein, Albert. Letter to Eduard "Tetel" Einstein, 1937.

PAGE 67
Bucky, Peter A. *The Private Albert Einstein*. Kansas City: Andrews and McMeel, 1992, 157.

Calaprice, Alice. *An Einstein Encyclopedia*. Princeton: Princeton University Press, 2015, 74.

Isaacson, Walter. *Einstein: His Life and Universe*. New York: Simon & Schuster, 2007, 358.

PAGE 73
Einstein, Albert. Letter to Michele Besso, January 5, 1929. Translator unknown.

PAGE 80
Viereck, George Sylvester. "What Life Means to Einstein: An Interview by George Sylvester Viereck." *Saturday Evening Post*, October 26, 1929.

INSERT BEFORE PAGE 81
Viereck, George Sylvester. "What Life Means to Einstein: An Interview by George Sylvester Viereck." *Saturday Evening Post*, October 26, 1929.

PAGE 84
Nelson, Anne. *Red Orchestra: The Story of the Berlin Underground and the Circle of Friends Who Resisted Hitler*. New York: Random House, 2009, 170.

INSERT BEFORE PAGE 85
Einstein, Albert. Submission to *Pax mundi: Libre d'or de la paix*, ca. 1930. Translator unknown.

PAGE 89
Ghose, Partha, ed. *Einstein, Tagore and the Nature of Reality*. New York: Routledge, 2017, 221.

Marianoff, Dimitri. "Einstein and Tagore Plumb the Truth." *New York Times Magazine*, August 10, 1930.

PAGE 90
Isaacson, Walter. *Einstein: His Life and Universe*. New York: Simon & Schuster, 2007, 376.

Viereck, George Sylvester. "What Life Means to Einstein: An Interview by George Sylvester Viereck." *Saturday Evening Post*, October 26, 1929.

PAGE 98
Einstein, Evelyn. "Foreword." In *Dear Professor Einstein: Albert Einstein's Letters to and from Children*. New York: Barnes & Noble Books, 2002, 13.

PAGE 103
Isaacson, Walter. *Einstein: His Life and Universe*. New York: Simon & Schuster, 2007, 345.

INSERT BEFORE PAGE 105
Einstein, Albert. *Albert Einstein: The Human Side*. Princeton: Princeton University Press, 1979, 18.

PAGE 117
Einstein, Albert, et al. "Can Quantum-Mechanical Description of Physical Reality Be Considered Complete?" *Physical Review* 47, no. 10 (1935).

PAGE 119
Klein, Martin J. "Thermodynamics in Einstein's Thought." *Science* 157, no. 3788 (August 4, 1967): 509

Schilpp, Paul Arthur, ed. "Autobiographical Notes." In *Albert Einstein, Philosopher-Scientist*. Evanston, IL: The Library of Living Philosophers, Inc., 1949, 33.

PAGE 122
Lenhart, Maria. "Lotte Jacobi: Absorbed in a Career of Extraordinary Images." *Christian Science Monitor*, February 24, 1981.

PAGE 124
Moriarty, Peter A. "Introduction." In *Lotte Jacobi: Photographs*. Boston: David R. Godine, 2002, 46.

INSERT BEFORE PAGE 125
Clark, Ronald W. *Einstein: The Life and Times*. New York: William Morrow, 2007.

PAGE 128
Nathan, Otto, ed. *Einstein on Peace*. New York: Simon and Schuster, 1960.

PAGE 131
Isaacson, Walter. *Einstein: His Life and Universe*. New York: Simon & Schuster, 2007, 131.

PAGE 132
Calaprice, Alice, ed. *The Expanded Quotable Einstein*. Princeton: Princeton University Press, 2000, 16.

PAGE 134
Warren, Lynne, ed. *Encyclopedia of Twentieth-Century Photography*. New York: Routledge, 2006, 1625.

PAGE 143
Pohlmann, Ulrich. "I Owe Everything to Landshoff." In *Hermann Landshoff: Portrait Mode Architektur*. Munich: Schirmer/Mosel, 2013.

PAGE 147
Richards, Alan Windsor. *Einstein as I Knew Him*. Princeton: Harvest House Press, 1979.

PAGE 148
Einstein, Albert. Letter to Marcus Campbell Goodall, September 10, 1945.

PAGE 151
Freer, Dawn. "Fred Stein (1909–1967): A Retrospective." In *"Escape to Life": German Intellectuals in New York; A Compendium on Exile after 1933*. Edited by Eckart Goebel and Sigrid Weigel. Boston: De Gruyter, 2012, 510.

PAGE 153
Calaprice, Alice, ed. *The Expanded Quotable Einstein*. Princeton: Princeton University Press, 288.

PAGE 156
Silverman, Lisa. "Trude Fleischmann." *The Shalvi/Hyman Encyclopedia of Jewish Women*, December 31, 1999. https://jwa.org/encyclopedia/article/fleischmann-trude.

PAGE 158
*Philippe Halsman: A Retrospective*. London: Pavilion Books, 1998, 64.

INSERT BEFORE PAGE 161
Einstein, Albert. Inscription on photograph by Yousuf Karsh, 1953.

PAGE 163
*Yousuf Karsh: Regarding Heroes*. Boston: David R. Godine, 2009, 162.

PAGE 164
Yeshiva University Libraries. "Credo." Accessed March 2, 2022. https://library.yu.edu/c.php?g=1073982&p=7880252.

PAGE 167
Casey-Maslen, Stuart. *The Treaty on the Prohibition of Nuclear Weapons: A Commentary*. Oxford: Oxford University Press, 2019, 5.

PAGE 172
Einstein, Albert. Letter to Marcel Sternberger, ca. 1950. Quote Courtesy of the Marcel Sternberger Collection, Copyright Stephan Loewentheil 2016.

PAGE 175
Isaacson, Walter. *Einstein: His Life and Universe*. New York: Simon & Schuster, 2007, 528.

PAGE 178
"Prof. Einstein Hails New Medical Center in Jerusalem." *JTA Daily News Bulletin*, June 5, 1952.

PAGE 184
This equation expresses Einstein's general theory of relativity.

PAGE 185
Howard, Don A. "Einstein's Philosophy of Science." *Stanford Encyclopedia of Philosophy*, revised September 13, 2019. https://plato.stanford.edu/entries/einstein-philscience/.

PAGE 189
Einstein, Albert. Letter to Dr. Mario Laserna Pinzón, January 8, 1955.

PAGE 192
Plaut, Frederick. *The Unguarded Moment: A Photographic Interpretation*. Englewood Cliffs, NJ: Prentice-Hall, 1964.

PAGE 196
"Dr. Albert Einstein Dies in Sleep at 76; World Mourns Loss of Great Scientist." *New York Times*, April 19, 1955.

INSERT BEFORE PAGE 197
Pais, Abraham. *Subtle Is the Lord: The Science and Life of Albert Einstein*. Oxford: Oxford University Press, 1982, 477.

PAGE 212
Isaacson, Walter. *Einstein: His Life and Universe*. New York: Simon & Schuster, 2007, 7.

# *INDEX*

*Page references in italics refer to illustrations.*

H



I



J



K



L



M



N

## *ACKNOWLEDGMENTS*

We are grateful to the following organizations and individuals who helped this book come to fruition:

The Antonia & Vladimer Kulaev Cultural Heritage Fund and the Sterling Foundation provided the financial support that made this book possible.

Tara Bandman, Myles DiRuggiero, and Oliver DiRuggiero contributed assistance and support throughout the project.

Greg Berger scanned and photographed the original documents.

William Bunch II gave expert advice on financing aspects of the project.

Yolanda Cuomo and Bobbie Richardson created the artistic design of the book and prepared it for printing.

Sjoerd Hofstra helped with translations from German to English.

Sally Knapp proofread the manuscript.

Karen B. Ringen, President of The Antonia & Vladimer Kulaev Cultural Heritage Fund, provided enthusiastic backing for the book.

Charles Tolkien-Gillett aided in research and fact-checking.

## *IMAGE CREDITS*

Cover and Pages 64, 150: Photograph by Fred Stein, © Fred Stein Archive.
Pages 123, 125, 127, 154: Courtesy of the University of New Hampshire.
Pages 133, 135–141: © The Magnes Collection of Jewish Art and Life.
Pages 142, 170: © Münchner Stadtmuseum, Sammlung Fotografie, Archiv Landshoff.
Page 159: Photo by Philippe Halsman © The Halsman Archive 2022.
Pages 161, 162–63: © Estate of Yousuf Karsh.
Page 173: Courtesy of the Marcel Sternberger Collection, Copyright Stephan Loewentheil 2016.
Page 196: Permission obtained from PARS International Corp.

*The images in this book are photographs or scans by Greg Berger of the prints and materials in the Berger collection. We have attempted diligently to trace the ownership of all copyrighted materials and secure the necessary permissions to be included in this publication. We apologize for and request to be notified of any inaccuracies or inadvertent omissions.*

*Einstein: The Man and His Mind*
Gary S. Berger, MD, and Michael DiRuggiero



Names: Berger, Gary S., author. | DiRuggiero, Michael, author.
Title: *Einstein: The Man and His Mind* / Gary S. Berger, MD, and Michael DiRuggiero.

Description: "Selections from the Berger Collection. Chapel Hill, North Carolina"—from title page.
Includes bibliographical references and index. | Bologna, Italy: Damiani, 2022.

Identifiers: ISBN: 9788862087841

Subjects: LCSH Einstein, Albert, 1879-1955. | Physicists—Biography.
BISAC BIOGRAPHY & AUTOBIOGRAPHY / Science & Technology

Classification: LCC QC16.E5 .B47 2022 | DDC 509—dc23



Printed in June 2022, Italy.

BOOK DESIGN BY YOLANDA CUOMO DESIGN

Associate Designer: Bobbie Richardson
Proofreader: Sally Knapp
Indexer: Cathy Dorsey
Retouching: Karina Eckmeier

Published by Damiani
info@damianieditore.com
www.damianieditore.com

*To punish me for my contempt for authority, fate made me an authority myself.*

A. Einstein